Swim Training Patterns

This pioneering book integrates mathematics into swim training to create an intellectual journey into patterns. Exercise your body and mind with over 35 training programs derived from mathematical ideas. Dive into the history of mathematics and computer science to discover structures that will enrich your exercise routine.

You don't need prior knowledge of mathematics or programming, just a curious mindset and the desire to swim interesting programs. This book will gently introduce you to the tools and knowledge you need to create programmatic training sessions. Learn how to write your training patterns using the Swimming Markup Language (swiML). Then, level up with the Python programming language to express even the most intricate training patterns. Creating swim training programs for every day of the week has never been easier.

Dr. Christoph Bartneck is a professor of computer science and a competitive swimmer with several national records. He actively promotes Masters Para Swimming in his role as national para swimming coordinator. He is an experienced science communicator with an interest in the intersection of mathematics, engineering and psychology. He frequently gives public talks and lectures at the local, national and international levels. Dr. Bartneck is an accomplished author with hundreds of scientific articles and books to his name. His work has been featured in the New Scientist, Scientific American, Popular Science, Wired, New York Times, The Times, BBC, Huffington Post, Washington Post, The Guardian, and The Economist.

Swim Training Patterns
Plan your Training Sessions with the Power of Mathematics

Christoph Bartneck, PhD

CRC Press
Taylor & Francis Group
Boca Raton London New York

CRC Press is an imprint of the
Taylor & Francis Group, an **informa** business

I APPROVE OF ANY ACTIVITY THAT REQUIRES
THE WEARING OF SPECIAL CLOTHING.
OSCAR WILDE

A DESIGNER KNOWS THAT HE HAS ACHIEVED PERFECTION
NOT WHEN THERE IS NOTHING LEFT TO ADD,
BUT WHEN THERE IS NOTHING LEFT TO TAKE AWAY.
ANTOINE DE SAINT-EXUPÉRY

...THE DESIGNER OF A NEW SYSTEM MUST NOT ONLY BE THE IMPLEMEN-
TOR AND THE FIRST LARGE-SCALE USER; THE DESIGNER SHOULD ALSO
WRITE THE FIRST USER MANUAL...IF I HAD NOT PARTICIPATED FULLY IN
ALL THESE ACTIVITIES, LITERALLY HUNDREDS OF IMPROVEMENTS WOULD
NEVER HAVE BEEN MADE, BECAUSE I WOULD NEVER HAVE THOUGHT OF
THEM OR PERCEIVED WHY THEY WERE IMPORTANT.
DONALD E. KNUTH

Contents

Acknowledgements

I would like to thank Peter Johnston and Kathryn Treeby for giving me valuable feedback on my writing. Callum Lockhart helped with the development of swiML. Christine McCurdy helped with her endless patience and critical thinking.

Acronyms

ASCII American Standard Code for Information Interchange. 11

CPU Central Processing Unit. 14

CSS Cascading Style Sheet. 195

CSV Comma–Separated Values. 3

DTP Desktop Publishing. 209

FINA Fédération internationale de natation. 220

FIT Interoperable Data Transfer Protocol. 13

GIF Graphics Interchange Format. 145

GPS Global Positioning System. 3

HTML Hypertext Markup Language. 12, 196, 207

IDE Integrated Development Environment. 191, 200

IOC International Olympic Committee. 13

IPTC International Press Telecommunications Council. 13

ISO International Organization for Standardization. 23, 215, 217

LZW Lempel, Ziv and Welch. 145

ODF Olympic Data Feed. 13

OWL Web Ontology Language. 10

PDF Portable Document Format. 22, 199

PyPI Python Package Index. 203

RDF Resource Description Framework. 10

Introduction

Summary

This chapter introduces the challenges that swimmers face when wanting to exercise their minds in addition to their bodies. It will also talk about how to approach the learning experiences in this book.

Swimming is one of the most popular endurance sports. Millions of amateur and a few professional swimmers swim up and down pools around the world. Swimming is a somewhat lonely activity since swimmers cannot talk to each other while swimming. They also do not directly interact with each other, unlike in games such as tennis. Swimming is also not a team sport, although there are relays. Most swimming is done as an individual sport. Swimmers compete against themselves as much as against others.

The hours that swimmers spend in the pool swimming endless laps of butterfly, backstroke, breaststroke and freestyle may seem strange to an outsider. "Is it not boring?" is a question that we are often confronted with. And while we do enjoy the activity itself, swimming certainly does not entertain as much as a volleyball match. Some swimmers, though, specifically enjoy the mindless physical activity, almost like a form of meditation.

Some swimmers have developed different ways to keep themselves mentally engaged. Humming melodies, counting laps, processing the events of the day, planning dinner, and focusing on breathing are all common mental strategies to deal with the inherent monotony of swim training.

Coaches generally make an effort to come up with new training sequences that vary in distance, stroke and equipment. One of the most frequently used swim training patterns is repetition. Swim 3×100m freestyle. Swim 8×50m drill. Sprint 4×25m butterfly with fins.

The Challenge

There are several problems with conventional swim training programs written on a whiteboard (see Figure 1). First, each coach develops his/her own language for writing the program, in particular when it comes to abbreviations. What does FT stand for? What is 8BB?[1] When swimmers

[1] FT is short for Finger Trail, a freestyle drill. 8BB stands for eight big breaths. This is a rest period that forces swimmers to focus on breathing and prevents them from chatting, a feature useful when coaching teenagers. Or very social swimmers.

Figure 1: A conventional swim training program.

[2] The irony of this sentence in the context of this book does not escape the author.

join a new club or try to find interesting new programs online, they often have to read a manual before using the program[2]. At times, the swimming instructions can also be ambiguous. Let's consider the example below in which K/D/S represent Kick, Drill, and Swim:

```
300 IM K/D/S
```

could be swum as:

```
100 IM Kick
100 IM Drill
100 IM Swim
```

or as:

```
75 Butterfly (25K/25D/25S)
75 Backstroke (25K/25D/25S)
75 Breaststroke (25K/25D/25S)
75 Freestyle (25K/25D/25S)
```

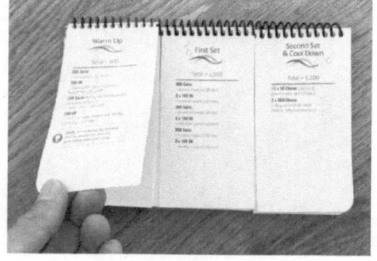

Figure 2: The Waterproof Swimmer book is subdivided into warm up, first/second set, and warm down.

Another problem is that conventional swimming programs cannot be read by computers. This may not bother you much, since computers are some of the least frequent swimmers. But if computers cannot read and understand a swimming program, then they cannot index them, which is essential if they are to be crawled by internet search engines such as Google. And if the search engines cannot process the training programs, then swimmers will be unable to find them. Wouldn't it be nice to be able to search for a 2,500-meter freestyle endurance program?

By far the biggest problem is the monotony of the programs. Denes (2018) tackled this problem by dividing his spiral-bound book into three sections (see Figure 2). Swimmers can combine sections to match their own preferences, resulting in a large number of possible combinations.

But it is still a book. Computers still cannot read and understand it. Furthermore, interested swimmers must first read the instructions for the book to comprehend the programs fully.

There are many sports technology products in the market which help athletes record their workouts. A Garmin or Apple watch can record your open water swim using its Global Positioning System (GPS) sensor (see Figure 10). The devices can even record your stroke. Indoors, they are fairly accurate at recording how many laps you swam and with which stroke. They sometimes struggle with recording your kicking since the watches are typically worn on your wrist and not on your ankle. The Form Smart Swim Goggles include a display that not only can show the swimmers what they have swum, but also what is next in the program. The display is limited to only a few characters per line, and hence, only a minimal amount of information can be displayed (see Figure 3). Taking a larger screen to the pool, such as a waterproof iPad, is only recommended when access to the pool is strictly controlled.

Figure 3: The Smart Swim Goggles by Form can display recent, current and planned swimming data.

These products need to exchange not only the swimmers' recorded data but also what the swimmers want to swim next. While the Form company offers workout planning with their membership program, they do not provide any documentation of their internal data format. They do allow an export to Garmin's FIT format (see page 13) or the open Comma-Separated Values (CSV) format. Neither data format is suitable for processing anything but the simplest swim training programs. They are, for example, unable to describe different drills or repetitions. They are also difficult, if not impossible, for humans to read or understand. For companies such as Form or MySwimPro, it is part of their business model to keep swimmers on their own platforms and to sell them membership subscriptions for training programs. Hence, opening their technologies to others is not in their best interest.

Solution

In this book, I will propose solutions for the problems mentioned above. First, I will describe a formal language for swim training programs, called Swimming Markup Language or swiML for short. Don't worry, we will find out what a markup language is in the next chapter.

Second, I will explore interesting patterns in mathematics and computer science that can be the basis for more intellectually fulfilling patterns in swim training programs. The goal is to allow you to exercise not only your body but also your mind.

Last, a technical description of swiML is provided to allow you to write your own programs with this language.

Bonus

You may ask: "Why not just use Generative Artificial Intelligence, such as the ChatGPT Large Language Model to create an endless number of interesting swim training programs?" I am glad you asked that question! Here is an excerpt from a chatGPT program:

ChatGPT ∨

Main Set (1500 meters)

1. **3 x 200 meters Freestyle (Descending Effort):**

    - 1st 200 meters at moderate pace (60% effort)
    - 2nd 200 meters at a faster pace (75% effort)
    - 3rd 200 meters at a fast pace (85-90% effort)
    - Rest 30 seconds between each 200 meters.

2. **6 x 100 meters (IM Order or Stroke Choice):**

    - Perform each 100 meters in the Individual Medley order (Butterfly, Backstroke, Breaststroke, Freestyle) or choose a mix of strokes.
    - Rest 15-20 seconds between each 100 meters.

3. **4 x 50 meters Sprint (Choice of Stroke):**

    - Perform these sprints at maximum effort (90-100% effort).
    - Rest 30 seconds between each 50 meters.

ChatGPT claims that the main program is 1,500 meters long but clearly it is only 1,400 meters. Large Language Models still suffer from hallucinations and fail at such simple tasks of adding up numbers. A hallucination in the area of Large Language Models is a response generated that contains false or misleading information presented as fact. This makes them fundamentally unreliable and therefore unsuitable for composing swim training programs.

Prerequisites

You don't need to know anything about programming or maths beyond what you might have learned in high school. As with physical exercise, the trick is to expose yourself to new challenges. You might not swim 100 meters freestyle in less than one minute on your first try. But if you exercise enough, this will become possible. The same holds true for intellectual challenges. You might not immediately understand every detail, but the challenge will exercise your mind.

Mathematics and programming are just languages to communicate ideas between humans and between humans and machines. Learning any new language will take effort but also yield a considerable reward. The Tools chapter on page 191 will introduce you to some of the tools that you can use to get started with the writing and programming of the swimming markup language.

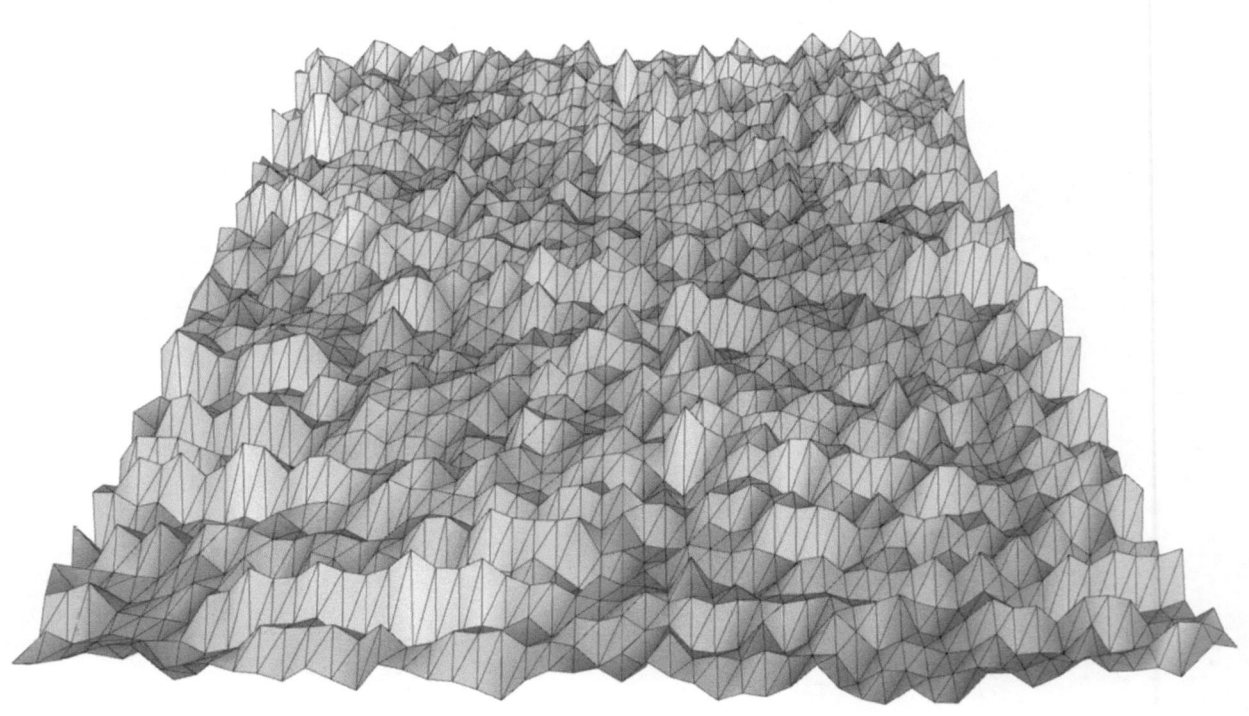

Language

Summary

This chapter discusses the foundations and challenges for natural and artificial languages – in particular the necessity for computers to understand language. Markup and programming languages are discussed to show how humans and computers can communicate with each other.

Swimmers and coaches worldwide write training programs using their own terms and grammar. There is considerable standardisation in competitive swimming. The strokes have their established names, such as freestyle, butterfly and individual medley. Fortunately, training programs offer more variety than swimming competitions, as coaches put effort into preserving the sanity of their athletes in the water who swim up and down the lanes thousands of times.

While their creative attempts to vary the swim training is greatly appreciated, it also introduces a communication challenge. A new drill to exercise a specific stroke requires a name and a description. These might differ from coach to coach and from swimmer to swimmer. There is no international organisation that ensures consistency and efficiency for the vocabulary and grammar of the swimming language in the way that the Académie Française does for the French language. But even if an organisation dedicates itself to maintaining a spoken language, it still cannot prevent ambiguities. Natural languages, such as English or French, are full of them. Coaches and swimmers have to deal with these ambiguities when talking about swimming. On the level of single words, there are several types of ambiguities:

homograph Words with the same spelling can have different meanings and different sounds. An example is "wind". It can be a gust of air or not following a straight path.

homophone Words sound the same but are spelled differently and have different meanings. Examples: "to", "two", "too".

polysemy One word can have different meanings. The word "mean" can *mean* "average" or "unkind".

synonym Words that are spelled differently and that sound different but have the same meaning. "Joyful" and "happy" are synonyms.

Ambiguities are not limited to single words. The order of words can also dramatically change the meaning. Consider these two sequences:

```
Free alcohol.
Alcohol free.
```

Which one would you prefer? When sequences of words are combined into a sentence, additional sources of confusion arise.

```
Visiting friends can be annoying.
```

What is the source of the annoyance? Friends who are visiting you or you visiting them? When the complexity of the sentence increases, the placement of a simple comma can again change the meaning as is entertainingly demonstrated in Lynne Truss's book "Eats, shoots & leaves" [3].

These ambiguities result in many misunderstandings in all aspects of our lives, including swimming. It is such a fundamental problem that great efforts have been made to find solutions. One of the most important challenges is to remove ambiguity from our language. One approach is to publish a brand new English dictionary and grammar text that resolves the problems. You would no longer be allowed to say "joyful". Instead, "happy" would be the only word to describe this concept. Publishing such a book might be possible, but changing people's habits might require an authoritarian dictatorship.

In his book 1984, George Orwell described such an authoritarian approach as "Newspeak" [4]. In his vision of the future, constraining the vocabulary had a far more sinister purpose. It was designed to prevent thought crimes and thereby control society. It would become impossible to even think about a crime since no words are left to describe it. The idea that language shapes thinking is often referred to as the Sapir-Whorf hypothesis based on the work of Edward Sapir and, in particular, his student Benjamin Lee Whorf [5]. The idea of changing and possibly improving human thinking through the use of language was picked up again in 1955 by James Cooke Brown with his Loglan language. This constructed language is designed to improve the speakers' logical thinking. By learning and speaking the language, far more precise and expressive logical thinking can be achieved.

It is unrealistic to expect that a natural language can ever be controlled in the same way a constructed language can. Even though the Académie Française tries to define the correct use of the French language, it cannot prevent English words from sneaking into usage. For example, many French would say "computer" rather than "ordinateur". Hence, natural languages will remain an organically developing system that is shared amongst its speakers. Any swimming language that is based directly on a natural language will always inherit its ambiguities.

Artificial Languages

Faced with the sheer insurmountable difficulties to rectify natural languages, it seems easier to start from scratch. This gives the designer of a newly constructed language the freedom to come up with new words and

[3] Lynne Truss. *Eats, Shoots & Leaves: The Zero Tolerance Approach to Punctuation.* Penguin, 2004. ISBN 9781861976123. URL https://search.worldcat.org/title/421779741

[4] George Orwell. *Nineteen Eighty-Four.* Secker and Warburg, 1949. ISBN 9780436350078. URL https://search.worldcat.org/title/12322711

[5] Paul Kay and Willett Kempton. What is the Sapir-Whorf hypothesis? *American Anthropologist*, 86(1):65–79, 1984. DOI: 10.1525/aa.1984.86.1.02a00050

grammar that prevent ambiguities. Esperanto is one of the most widely spoken constructed languages. Widely spoken in this context means more spoken than Elvish or Klingon, two artificial languages based on fictional stories.

> **Bonus**
>
> Ludwik Zamenhof (see Figure 4) first published the concept of Esperanto in his book "Dr. Esperanto's International Language" in 1887. It contains vocabulary and grammar, similar to other foreign language guides, and was intended to improve international communication. Today, several thousand people actively speak Esperanto, and you can even enjoy Tolkien's "The Hobbit" in Esperanto. This group of Esperanto enthusiasts form a social community that even has its own flag (see Figure 5).

Figure 4: Ludwik Lejzer Zamenhof (15 December 1859 – 14 April 1917) invented Esperanto.

In Esperanto "swim 100 meters freestyle" translates to:

`Naĝu 100 metrojn liberstila`

It would be possible to create an entirely new language to describe swimming. Besides the clear disadvantage of having to learn a whole new language to be able to follow swimming instructions, it does not necessarily resolve all the logical ambiguities in training instructions. If the underlying concepts are unclear, then the words used to describe this concept will remain ambiguous. Still, creating a perfectly structured language to define meaning is necessary to overcome ambiguity. This necessity is one of the major obstacles to developing software.

Figure 5: The flag of the Esperanto movement.

Ontology

Whenever new software is created, the developers have to represent a small slice of reality in software. A computer knows by default nothing about the world, and any task it is to do needs to be unambiguously explained to it. A computer has no common sense, instinct or experiences. Its brain is fundamentally empty. It also cannot understand natural language. You cannot easily just tell it what to do, although some progress has been made in recent years. Your smartphone's voice assistant, such as Siri, can respond to a limited number of verbal commands. Still, we are a long way away from being able to program a computer by using natural language, whether spoken or typed.

Let's take a short and easy example of a new pool management software. The developers have to define the concepts of "pool", "customer", "invoice", etc. For each customer, attributes such as first name, last name, address, and phone number need to be defined. The formal specification of the concepts and their relationships that exist within a particular domain is called an *ontology*. It provides a way to define a shared vocabulary that can be used to describe and reason about entities, such as objects, concepts, and events, within a given domain. Such an ontology is similar to an artificial language. It does not create new words, but it explicitly defines the meanings for each word used.

Figure 6: The World Wide Web Consortium develops and maintains OWL as part of its Semantic Web activity and its Data activity:
https://www.w3.org/TR/owl-features/

And of course there is a formal standard for describing such an ontology: The Web Ontology Language (OWL) and the Resource Description Framework (RDF). They were developed by the World Wide Web Consortium (W3C). They use nouns representing classes of objects and verbs representing relations between the objects. This representation is somewhat similar to the structures in object-oriented programming, but we will not go into this any deeper now.

An ontology defines unique and unambiguous vocabularies. In our example, the concept of a person who buys admission to a pool is called "customer". Not "shopper", "buyer", or "client". His/her name is recorded with "first name" and "last name" and not "given name" and "family name". We also define that each customer can have only one last name but several first names. Moreover, each customer can purchase access to one or many pools, such as the main sports pool and the extra spa pool.

The software for the pool will be built upon the ontology developed. Users of the software complete their daily tasks with this simplified model of reality. Reality is complex and chaotic, and no ontology can model it all while still being easy to implement in software. There will always be some specific exceptional cases that cannot be captured in a particular ontology. For example, a new policy might come into action that allows disabled swimmers to take a support person with them for free. Now we have a customer that enters the pool without paying for it. If the development team was smart, then they might have implemented some exceptional type of customers, such as an "entrance free" option to capture all the exceptional reasons why a person might get admission for free without specifying the reason. But such foresight cannot be guaranteed. We'd better stop here before we define a whole swimming pool platform. It is only important to point out that any ontology and software based on it are simplified models of reality.

To formally describe the swimming pool domain, or any domain for that matter, OWL and RDF use a tag-based approach. Let's consider a short example for a customer. A customer's data can be captured as:

```
<customer>
  <firstName>Michael<\firstName>
  <lastName>Phelps<\lastName>
<\customer>
```

Here we have `<tags>` that are enclosed by the less than < and the greater than > symbol. The < and > in this context do not mean greater or lesser, but they signify the start and end of a tag. Furthermore, each tag needs to be opened and closed. Closing a tag is done by adding the backslash \ symbol before the tag name `<\tagName>`. Enclosed by the tags is the data. This can be the actual value, such as "Michael", or other tags.

Notice that the `<customer>` tag includes two other tags. It includes `<firstName>` and `<lastName>`. In the example above, we tell the computer that this customer has the first name "Michael" and the last name "Phelps". The vocabulary of our swimming pool ontology needs to be carefully constrained. There can only be one tag `<firstName>`. There cannot be another tag `<givenName>` that describes the name of the customer. By constraining the vocabulary used for tags and data we can avoid ambiguities. We can also define the relationships between concepts. A customer has a first and last name.

The main conceptual step forward here is the pairing of a tag with data. This could in principle also be achieved in other ways, such as `firstName=Michael`. Still, the idea of having a strictly controlled vocabulary for tags that describe the nature of the data is essential. It is a good compromise for a language that can be read by humans *and* computers.

Notice that the tags here exclude spaces. It is `<firstName>` and not `<first name>`. The reason is again to avoid ambiguity. Computers know several different types of invisible characters that are only perceivable by the spacing they create between words and characters. And of course we have another standard to describe all the characters, symbols and emojis used in all languages called Unicode. It is maintained by the Unicode Consortium. The unique identification of the `A` character, for example, is `U+0041`. The swimming emoji is `U+1F3CA` (see Figure 7).

Bonus

Defining a character, such as `A`, by using another sequence of characters, such as `U+0041` does seem like a circular argument. We need to dive a little bit deeper into how computers make sense of characters. They don't. Computers can only understand numbers. Two at that. Zero and one.

In 1963 the American Standards Association published the American Standard Code for Information Interchange (ASCII). It gave each character a number. `A` was given the number 65. ASCII only included 128 characters since this could be stored in seven bits of memory. It did not have enough space to include characters from non-English languages, such as the German umlaut. Not to mention non-Latin based languages, such as Japanese or Arabic. Workarounds were implemented for a while, but it became clear that a much larger database for all the characters of all languages was necessary. In 1993 the first version of Unicode was adopted which was backward compatible with ASCII. The letter `A` still had the number 65.

Numbers can be expressed in different ways. On the lowest level, computers only understand zero and one. The number 65 in the binary numeral system is `1000001`. But it can also be expressed in the hexadecimal numeral system. You will learn more about these numeral systems in the section Binary on page 63. For now, you need to be content with a short example. The hexadecimal numeral system has a base of 16. Counting from 0 to 18 is:

`0,1,2,3,4,5,6,7,8,9,A,B,C,D,E,F,10,11.`

Notice that 10 in hexadecimal is equal to 16 in decimal. The decimal number 65 is 41 in hexadecimal. `U+` only indicates that the following is a Unicode. The next two numbers are simply leading zeros. 41 is then the hexadecimal number for the letter `A`. We will revisit binary numbers in a training pattern on page 63.

Figure 7: The Unicode Consortium maintains a database of all the characters used in all the languages, including the emojis on your phone. It currently includes 149,186 characters, covering 161 modern and historic scripts. It also includes 3664 emoji, including several for swimming. The swimming emoji is `U+1F3CA`. It is here shown in the Noto Typeface, which was created by Google, but is available under the SIL Open Font License. The goal of the Noto typeface family is to cover all of the Unicode characters (`https://notofonts.github.io`).

We have the classic white space symbol that you get when hitting the space bar ☐ on your keyboard with the Unicode U+0020. When you press the tab → key, you get the Character Tabulation space (Unicode: U+0009). There is also the Non-breaking Space (Unicode: U+00A0) which is similar to the classic space, but it will not show any space if it occurs at the end of a line. In Microsoft Word, you can access this character by the keyboard shortcut ctrl + ⇧ + ☐. There is even a zero-width space character (Unicode: U+200B), which is used in languages that do not typically use spaces between words, such as Japanese.

There are several strategies to visually separate words without using white space. Above, the camelCase system was used in which all words are concatenated. The first character of each new word is capitalised: firstName. You will need to use your visual imagination to understand why this convention refers to a camel. Other conventions are to replace the space character with less ambiguous characters, such as the underscore character first_name or the dash character first-name. All of these conventions are used in the World Wide Web. There, each web page has a unique address called Universal Resource Locator (URL). No space characters are allowed in a URL. Wikipedia's article on the swimming sport named "Swimming (sport)" has the URL:

https://en.wikipedia.org/wiki/Swimming_(sport)

The space character in the title was replaced with an underscore _ character in the URL.

The concept of using a combination of tags and text to describe knowledge is the underlying principle for many markup languages, including the Hypertext Markup Language (HTML). HTML is the most widely used markup language, and it is maintained by the W3C. It describes the structure and layout of all web pages. In HTML the code:

```
<p>This is an <em>italic</em> word</p>.
```

Is displayed by a web browser as:

This is an *italic* word.

<p> is the tag for a paragraph and <em> is used to for the italic style, also called *emphasis*. HTML has a fixed number of standardised tags. But what if you want to describe your own knowledge domain? One of the most frequently used tag-based languages to describe knowledge is the Extensible Markup Language (XML) which is also managed by the W3C (see Figure 8). It allows developers to define their own tags, such as <firstName>. The XML language and its tag system are the basis for describing many knowledge domains. For example, it includes BeerXML, a language to exchange brewing information, and MusicXML, a language for musical notations, to name just a few.

One of its great advantages is that it allows developers to define a schema for their markup language. Schemas precisely define what is and what is not allowed in a language. A schema not only defines the tags but also their relationships. A schema could define that each customer can only have one last name. The last name could be constrained from using any

Figure 8: The World Wide Web Consortium develops and maintains the XML standard.

numbers or special symbols. A document can then be validated against such a schema.

Markup Languages

There are several XML based markup languages that describe the sports domain. The International Olympic Committee (IOC) created an XML based markup language for their competitions called the Olympic Data Feed (ODF)[6]. It is used to automatically communicate the results of competitions, such as at the Olympic Games, between the local venues and the central administration. This flow of information is also available to external organisations, such as news agencies. They can use it to automatically receive the latest results. It was first used at the 2010 Winter Olympics in Vancouver. For swimming it includes, amongst others, information about the athletes, the races and the results.

[6]https://odf.olympictech.org

Sports is an important segment in the news, and it is not limited to the events organised by the IOC. Every day, news agencies report on soccer, volleyball and cricket matches, just to name a few. Writing the results of sports news has become an almost automatic process using modern software. The International Press Telecommunications Council (IPTC) created the SportsML[7] XML based standard to facilitate communication. Many major news agencies, such as AP, BBC and Reuters use this standard. SportsML covers many major sports, but it lacks a specific standard for swimming. The IPTC does have a controlled vocabulary that is used to describe the topics of a news item. This set of NewsCodes includes terms used in swimming, such as the names of the strokes.

[7]https://iptc.org/standards/sportsml-g2/

Both of these XML markup languages focus on competitions. They cannot describe training sessions and are therefore of limited use in describing swimming patterns. There are other XML based markup languages to describe sports training. Garmin, one of the major sports technology companies (see Figure 9), developed several such languages over the years.

In 2007 they introduced the Training Center XML (TCX) standard that enabled athletes to exchange data that was recorded with one of Garmin's many products, such as GPS sport watches and heart rate monitors. Garmin openly shares this standard that is based upon XML.

Figure 9: Garmin is one of the major sports technology companies who developed two XML based training languages.

> **Bonus**
>
> Garmin's TCX standard is currently available on their website. It is unclear how long they will remain openly available, and hence I stored these important documents in an online repository:
> https://github.com/bartneck/swiML/tree/main/garmin

In 2010 Garmin introduced the Interoperable Data Transfer Protocol (FIT) protocol[8]. It does not use XML and is hence not human-readable. Instead, it focuses on defining a technical messaging system. It does, for example, enable a cadence sensor attached to the cranks of your bicycle to communicate wirelessly with your cycling computer. The cycling computer can then communicate with your smartphone or desktop computer. Garmin documents this messaging protocol openly, and it has been picked up by

[8]https://developer.garmin.com/fit/protocol/

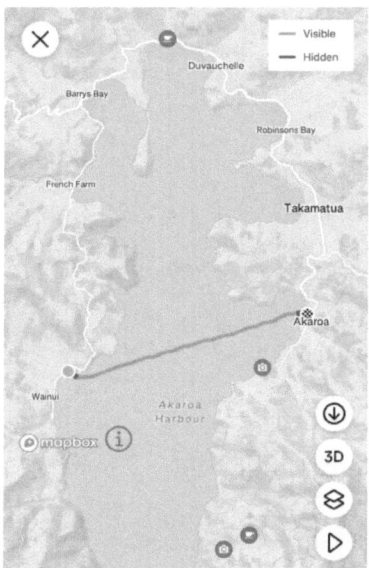

Figure 10: GPS is useful for open water swimming, but less so for swimming in a pool, even if is outdoors.

other sports technology companies, such as Polar, Suunto, Wahoo and Strava. Both of these standards focus on recording the performance of the athlete, and they have little to no facility to describe a planned workout.

These two markup languages and protocols are designed to record the actions of the athlete. They can combine a variety of sensory information. For cyclists, their GPS location, altitude, heart rate, cadence, power on the pedal and speed at the wheel can be recorded. In addition, wind direction, temperature, humidity, and other environmental factors can be supplemented through online services. All of this data needs to be stored in the data file, and it is great that we have modern formats that can do this. For swimming, GPS is only useful for open water events (see Figure 10).

A training markup language, however, has a slightly different focus. It does not record what you have done, but what you should do. It is a list of instructions rather than a log. Hence a swimming markup language must include instructions that tell the swimmer what to do next. At its heart, it is a small programming language. Instead of instructing the computer to add two numbers, it instructs swimmers to execute certain exercises. Moreover, this cannot be a simple list of instructions that could be stored in a table. It must include concepts that influence the sequence of instructions. For example, it must include the concept of repetitions. It must be possible to express 4×100 `Freestyle`. In computer programming, this concept is called a *loop*.

Computers use already different languages for their operation. At the lowest level, they use machine language, which can be directly executed by the Central Processing Unit (CPU). This machine language is written in the binary numeral system. It consists of sequences of ones and zeros:

`01001010 10101000 01000111 10111111`

Machine language is extremely difficult for humans to read and understand. The Assembler language (see Listing 1) was developed to make it slightly easier to write a program. It is still similar to machine language, and it can quickly be assembled into machine language.

```
1      section .data
2  section .text
3      global _start
4  _start:
5      mov eax, 42
6      add eax, 8
7      mov ebx, eax
8      mov eax, 1
9      xor ecx, e
```

Listing 1: This assembler program sets the EAX register to 42, adds 8 to it, stores the result in EBX, sets EAX to 1, clears ECX, and then invokes a system call to exit the program. Note that this is a very basic example and doesn't include any input or output operations.

If you understand this program, then you are a computer scientist and a good one at that. This low-level programming language is still hard to learn, understand and use. Hence, higher programming languages were developed that were much easier for humans to understand. These higher programming languages need to be compiled into machine code before they can be executed. Here is a short example in the C language (see Listing 2):

```
1   int main() {
2       int a = 10, b = 5;
3       int sum = a + b;
4       printf("The sum of %d and %d is %d\n", a, b, sum);
5       return 0;
```

Listing 2: This C program adds two numbers and prints the result on the screen.

The programming language C is already considered a higher programming language, but modern languages have become even easier for humans to use and understand.

Python (see Figure 11) is currently considered one of the easiest to use languages. The same program in Python would look like (see Listing 3):

```
1   a = 10
2   b = 5
3   sum = a + b
4   print("The sum of", a, "and", b, "is", sum)
```

Figure 11: The Python programming language was first released in 1991 by Guido van Rossum.

Listing 3: This Python program does the same as Listing 2.

All of these higher programming languages have a restricted vocabulary for their commands. In Listing 3 the term print is a reserved word that prints text to the screen. We also encounter data in the form of variables and their values. a is a variable with the value 10. We will not go any deeper into Python at this point, but it is important to notice that programming languages are artificial languages that have a constrained vocabulary and grammar. Just like the markup languages or constructed languages we discussed above. We could also express the program above in plain English:

INPUT: What is the sum of 10 and 5?
OUTPUT: The sum of 10 and 5 is 15.

Or in Esperanto:

ENIGAĴOJ: Kio estas la sumo de 10 kaj 5?
ELIGOJ: La sumo de 10 kaj 5 estas 15.

It can also be expressed in the language of Maths:

$$10 + 5 = 15 \tag{1}$$

Again, we have a constrained vocabulary of words, such as $+$ meaning addition and $=$ meaning equal. It also knows data in the form of numbers 5 and 10.

It is important to find the right balance between a language being readable for humans and readable for computers. An XML based markup language is this middle ground. It can be read, validated and imported by computers while still being readable for humans. For an XML based markup language to be successful, it has to satisfy several criteria:

concise Being able to express the desired concept in an efficient way.

expressive Being able to express all desired concepts effectively.

flexible Being able to adapt the language to suit the application domain

constrained Being rigid enough to avoid ambiguity.

open Everybody should have free access to the language.

learnable It must be easy to learn the language.

In the next chapter, I will introduce the Swimming Markup Language (swiML), which is an XML-based language to describe swim training programs. swiML has been designed with these criteria in mind, although its development will continue. There is always room for improvement, and several extensions have already been identified. Feel free to make a feature request at:
https://github.com/bartneck/swiML/issues.

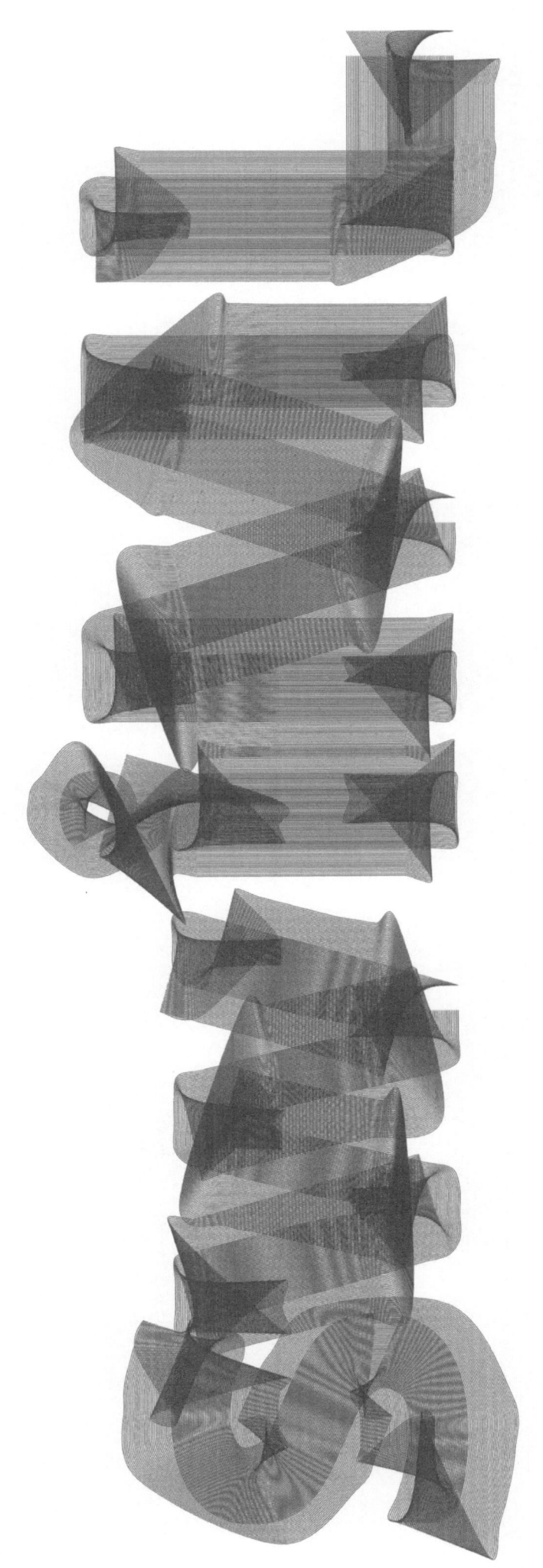

Swimming Markup Language

Summary

This chapter introduces the Swimming Markup Language. It shows how to compose a simple swim training program and how to transform and validate it. Essential elements are discussed, including structural and directional elements.

The Swimming Markup Language (swiML) is designed to describe swim training sessions (see Figure 12). It is based on the XML language. It is available free online[9], and it is the first standard that enables coaches, swimmers, and technology companies to exchange training programs. This chapter provides a short introduction to this language while the full documentation is available on page 207. Let's start with a simple swim training program:

[9]https://github.com/bartneck/swiML

Figure 12: The logo of the Swimming Markup Language.

```
1   <?xml version="1.0" encoding="UTF-8"?>
2   <program>
3       <poolLength>25</poolLength>
4       <lengthUnit>meters</lengthUnit>
5       <instruction>
6           <length>
7               <lengthAsDistance>100</lengthAsDistance>
8           </length>
9           <stroke>
10              <standardStroke>freestyle</standardStroke>
11          </stroke>
12      </instruction>
13  </program>
```

Listing 4: A simple swim program in swiML.

This simple training program only consists of 100 meters freestyle. Twelve lines of code do seem a lot for such a short program, but its elements will quickly become more useful. Authoring tools described on page 191 dramatically speed up the process of writing swiML programs. This swiML program is not intended for the swimmer at the pool. They will see a transformation as shown in Figure 16, and we will learn how to do this transformation shortly. For now, let's focus on the XML code itself.

The first line in Listing 4 declares that this program is written in XML version 1.0, and its characters are encoded in UTF-8. Unicode Transformation Format (UTF) is a character encoding standard using Unicode, which we already discussed on page 11. The number 8 refers to the 8-bit version of UTF. This first line informs any software that intends to use it what to expect of it.

Line 2 opens the element `<program>`, which is the root element for the whole document. Line 3 defines the pool length for which this program is designed, and line 4 defines the distance unit. While most pools are likely to be 25 meters or 50 meters, there are some pools that use yards, and on rare occasions, fractional distances, like 33.3 meters. Specifying the length and unit of the pool is necessary to allow declaring the length of instructions in laps. Without it, it would be unclear how many meters two laps would be, for example. Line 5 opens the `<instruction>` element. Each `<program>` element can contain one or many `<instruction>` elements.

The `<instruction>` must have at least a length and a stroke. In this example, length is defined as `<lengthAsDistance>`, which uses the `<lengthUnit>` as its unit. In this case, this is meters. Since the `<poolLength>` and `<lengthUnit>` are already defined, it would also be possible to define the length as `<lengthAsLaps>`. 100 meters would require 4 laps. Line 9 then declares the `<stroke>` for the 100 meters. In this case, it is the `<standardStroke>` freestyle.

Validation

This XML file can be read by a computer and validated against the swiML schema. The schema defines the constraint vocabulary and grammar for this markup language. We have to tell the computer which schema our swim program should comply with. For this, we have to extend the `<program>` element with several attributes:

```
1   <?xml version="1.0" encoding="UTF-8"?>
2   <program xmlns="https://github.com/bartneck/swiML"
3       xmlns:xsi="http://www.w3.org/2001/XMLSchema-instance"
4       xsi:schemaLocation="https://github.com/bartneck/swiML
        ↪  https://raw.githubusercontent.com/bartneck/swiML/main/version/
        ↪  latest/swiML.xsd">
5       <poolLength>25</poolLength>
6       <lengthUnit>meters</lengthUnit>
7       <instruction>
8           <length>
9               <lengthAsDistance>100</lengthAsDistance>
10          </length>
11          <stroke>
12              <standardStroke>freestyle</standardStroke>
13          </stroke>
14      </instruction>
15  </program>
```

Listing 5: A swiML program with a name space and schema location.

The `xsi:schemaLocation` attribute points to the URL of the swiML XML Schema Definition (XSD) file, which is hosted on the GitHub site. GitHub is a popular hosting service for software developers (see Figure 13). It allows them to share code and keep track of the changes to the code. You can download and modify all the swiML software from there[10]. The `swiML.xsd` file defines what a valid swiML file is. A more detailed discussion of the swiML is available on page 207.

The `<program>` element includes several other important attributes. First, there is the XML Name Space (XMLNS) attribute. Name spaces define the context for an XML language. The `<program>` element in swiML refers to a swim training set. In the context of a music concert, it could refer to the sequence of songs performed. All the other ambiguities for words discussed on page 7 apply, such as polysemy, homographs and synonyms. To avoid element ambiguity, it is necessary to give each XML language a unique name space. This could be any sequence of characters as long as it is unique. swiML uses the GitHub URL `https://github.com/bartneck/swiML` as its name space since by definition each URL has to be unique. URLs are the addresses for web pages, and the World Wide Web would stop functioning properly if web pages didn't have unique addresses. The `xmlns:xsi` attribute defines a name space prefix which we will not need to discuss here.

Figure 13: Github is a popular code sharing platform.

Metadata

The attributes of the `<program>` element include information about the information. This is often referred to as metadata. So far, the metadata has focused on the technical aspects of the swiML language, such as its name space. It is also possible to include important metadata about the training set. This includes a title, author, creation date, and description (see Listing 6). It can also contain global parameters, such as the width of the program layout or whether the metadata should be displayed at all. But more on that later.

```xml
<?xml version="1.0" encoding="UTF-8"?>
<program xmlns="https://github.com/bartneck/swiML"
     xmlns:xsi="http://www.w3.org/2001/XMLSchema-instance"
     xsi:schemaLocation="https://github.com/bartneck/swiML
     ↪ https://raw.githubusercontent.com/bartneck/swiML/main/version/
     ↪ latest/swiML.xsd">
    <title>Jasi Masters</title>
    <author>
        <firstName>Christoph</firstName>
        <lastName>Bartneck</lastName>
    </author>
    <programDescription>Our Tuesday evening program in the sun. The
     ↪ target duration was 60 minutes.</programDescription>
    <creationDate>2023-02-07</creationDate>
    <poolLength>25</poolLength>
    <lengthUnit>meters</lengthUnit>
    <hideIntro>true</hideIntro>
    <instruction>
        <length>
            <lengthAsDistance>100</lengthAsDistance>
        </length>
        <stroke>
            <standardStroke>freestyle</standardStroke>
```

Figure 14: The Adobe company created the Portable Document format in 1992.

[11] https://github.com/bartneck/swiML/blob/main/swiML.xsl

Figure 15: For this HTML version of the program the `<hideIntro>` element was set to false.

```
 21          </stroke>
 22        </instruction>
 23    </program>
```

Listing 6: A swiML program with metadata.

Many of the meta elements, such as `<title>` and `<creationDate>` are self-explanatory. The element `<hideIntro>`, however, requires a bit more attention. The swiML file can be read and understood by both humans and computers. Still, it is verbose and not suitable for writing on a whiteboard at the pool. It would take up too much space, and the readability could be improved. The XML file needs to be transformed, and the role that the `<hideIntro>` element plays will soon become clearer.

Transformation

Since swiML is well defined through its schema (XSD file), a computer can transform this XML data to other representations that are easier for humans to read. The Extensible Stylesheet Language Transformations (XSLT) is a programming language designed for this purpose. The swiML project contains an XSLT program to transform the XML to HTML[11]. This HTML can be viewed with any web browser (see Figure 15). It can also be converted to the Portable Document Format (PDF). PDF is a language to describe the content on a page, although it is not directly based on XML. PDF is commonly used for documents that are intended for printing (see Figure 14).

Jasi Masters

Christoph Bartneck
Our Tuesday evening program in the sun. The target duration was 60 minutes.

⤳Date:30 July 2023
⤳Pool Size:25
⤳Units:meters
⤳Length:100 meters / 4 Laps

100 FR	1

made with: **swiML**

The element `<hideIntro>` can be used to exclude the metadata from the HTML representation. If it is set to `true`, then the XSLT transformation excludes all the metadata (see Figure 16).

```
100 FR                                                     1
```

Humans certainly find it easier to read this HTML representation compared to the XML representation.

> **Bonus**
>
> Both swiML output formats, HTML and PDF, use the free and open source JetBrains Mono typeface `https://www.jetbrains.com/lp/mono/`. It is a mono-spaced typeface, meaning that all characters have the same width. The letter i has the same width as the letter m. Moreover, all characters have a distinct shape. The letter O can easily be distinguished from the number 0. Moreover, the letters l, I and the number 1 are recognisable, which cannot be said for the many other typefaces: l,I,1. The JetBrains Mono typeface is easy to read and hence best suited for bringing swim training instruction to the pool. Even with foggy goggles or at a distance, you will be able to read the program. All `code` in this book also use this typeface.

Essential Instructions

Each `<program>` can have one or many `<instruction>` elements. Each `<instruction>` must have at least a length and a stroke. Length can be provided as `<lengthAsDistance>`, which would refer to meters in the case above. But it is also possible to define lengths through `<lengthAsLaps>` or `<lengthAsTime>`. The stroke can be defined as `<drill>`, `<kicking>` or `<standardStroke>`. A `<standardStroke>` can be, for example, freestyle or individual medley.

The length and stroke elements are required for an instruction to be valid. However, there are many optional elements that can better describe an instruction.

First, each `<instruction>` can have a `<rest>` element. This defines how much rest the swimmer has after completing the length. `<rest>` can be defined as, for example, the period since the swimmer started to swim. This is expressed as `<sinceStart>`. The Listing 7 shows this in context. It means that the swimmers have to start swimming 1:45 minutes after they started to swim the 100 meters freestyle. The full details of all rest options are available on 217.

The duration of the rest is mostly given in a time format. XML has several basic data types, and duration is one of them. It is expressed in the ISO 8601 standard under the form `PnYnMnDTnHnMnS` which is maintained through the International Organization for Standardization (ISO) (see Figure 17). Our instruction would look like:

Figure 17: The International Organization for Standardization manages the ISO 8601 standard since 1988.

```
1   <instruction>
2       <length>
```

```
3          <lengthAsDistance>100</lengthAsDistance>
4       </length>
5       <stroke>
6          <standardStroke>freestyle</standardStroke>
7       </stroke>
8       <rest>
9          <sinceStart>PT1M45S</sinceStart>
10      </rest>
11   </instruction>
```

Listing 7: A swiML program with a 1:45 rest duration.

Bonus

ISO 8601 defines how dates and times are written as text. The format used for the duration is PnYnMnDTnHnMnS. P simply means "period" and is followed by n number of years nY, months nY and days nY. In the context of swimming, we are unlikely to encounter a swimming duration that is one day or longer. Hence we can move forward to T which denotes time. Here we can define n number of hours nH, minutes nM and seconds nS. Two minutes and thirty seconds would be PT2M30S.

Besides the rest, `<instruction>` can also include the `<intensity>`. It describes the intensity at which an instruction should be swum. This can be achieved by declaring the `<startIntensity>`. It defines that the swimmer should swim at a constant intensity. A simple intensity instruction could be:

```
1    <instruction>
2       <length>
3          <lengthAsDistance>100</lengthAsDistance>
4       </length>
5       <stroke>
6          <standardStroke>freestyle</standardStroke>
7       </stroke>
8       <intensity>
9          <startIntensity>
10            <zone>easy</zone>
11         </startIntensity>
12      </intensity>
13   </instruction>
```

Listing 8: A swiML program with a static intensity.

Repetitions

Computers are endlessly patient, and hence, they excel at repetitive tasks. Loops are one of the pillars of most programming languages. swiML includes the concept of repetitions, which is also a common feature in many training programs. To model four times 100 meter freestyle we can simply add a `<repetition>` element that in turn includes a `<repetitionCount>` element. A repetition then includes one or more `<instruction>` elements (see Listing 9).

```
1   <instruction>
2       <repetition>
3           <repetitionCount>4</repetitionCount>
4           <instruction>
5               <length>
6                   <lengthAsDistance>100</lengthAsDistance>
7               </length>
8               <stroke>
9                   <standardStroke>freestyle</standardStroke>
10              </stroke>
11          </instruction>
12      </repetition>
13  </instruction>
```

Listing 9: A swiML program with a repetition.

An `<instruction>` element in a `<repetition>` can include more `<repetition>` elements, and hence this nesting can continue indefinitely. This concept of an element or function referring to itself is called recursion.

```
1   <instruction>
2       <repetition>
3           <repetitionCount>4</repetitionCount>
4           <instruction>
5               <length>
6                   <lengthAsDistance>100</lengthAsDistance>
7               </length>
8               <stroke>
9                   <standardStroke>freestyle</standardStroke>
10              </stroke>
11          </instruction>
12          <instruction>
13              <repetition>
14                  <repetitionCount>2</repetitionCount>
15                  <instruction>
16                      <length>
17                          <lengthAsDistance>50</lengthAsDistance>
18                      </length>
19                      <stroke>
20                          <standardStroke>backstroke</standardStroke>
21                      </stroke>
22                  </instruction>
23                  <instruction>
24                      <length>
25                          <lengthAsDistance>50</lengthAsDistance>
26                      </length>
27                      <stroke>
28                          <standardStroke>breaststroke</standardStroke>
29                      </stroke>
30                  </instruction>
```

```
31              </repetition>
32            </instruction>
33        </repetition>
34  </instruction>
```

Listing 10: A swiML program with a nested repetition.

The program in Listing 10 is rendered in HTML using parentheses (see Figure 18). Notice that the HTML conversion of swiML includes light grey line numbers on the right. They help swimmers to refer to a specific line of instruction.

Figure 18: This HTML rendering shows the <instruction> defined in Listing 10.

$$
4 \times \left[\begin{array}{l} \textbf{100} \ \text{FR} \\ 2 \times \left[\begin{array}{l} \textbf{50} \ \text{BK} \\ \textbf{50} \ \text{BR} \end{array} \right. \end{array} \right.
$$

1
2
3

> **Bonus**
>
> Recursion is one of the methods in programming that puzzles many beginners and occasionally also seasoned developers. Many mistake a recursion for a repetition or loop. Recursion works by using a function that can call itself for the solution. It thereby solves the problem by first solving smaller instances of the same problem. Typically problems that can be easily solved by recursion are finding the factorial of a number or calculating a Fibonacci number (see page 71).
>
> The Google company does have some sense of humour. When you search for the term "recursion" Google asks "Did you mean: recursion", which is a hyperlink to the search for the term "recursion". Normally, Google only shows suggestions when you misspell a word. With the exception of "recursion". Try it yourself with this URL: `https://www.google.com/search?q=recursion.`
>
>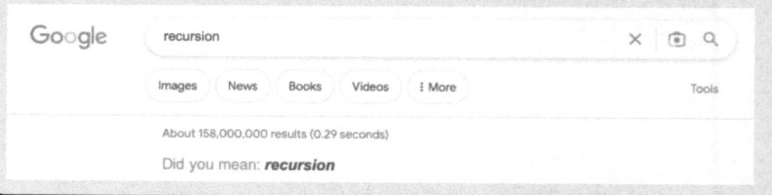

Extended Instructions

Instruction elements can have many more optional modifications.

Breathing

Breathing is a necessary but slightly inconvenient activity for swimming. Breathing on every stroke slows swimmers, and hence swimmers train their respiratory system. In particular for freestyle it is common to use alternating breathing, such as breathing on every 3rd or 5th stroke. The coach can use the `<breath>` element to instruct swimmers to breathe on a specific number of strokes.

Equipment

Swimmers use a variety of equipment in their training. Adding this to swiML is easy. Just add the `<equipment>` element to the `<instruction>` element. The currently supported types of equipment include pads and fins.

Description

The creativity of coaches and swimmers is limitless. Coming up with a formal description for all possible aspects of training is a Sisyphean task. Hence swiML uses the `<instructionDescription>` to add a free description to the `<instruction>` element. This enables coaches to describe even exotic exercises, such as balancing a paddle on your head while swimming. It can be written as shown in Listing 11:

```
1   <instruction>
2       <length>
3           <lengthAsDistance>100</lengthAsDistance>
4       </length>
5       <stroke>
6           <standardStroke>freestyle</standardStroke>
7       </stroke>
8       <instructionDescription>Paddle on your
    ↪   forehead</instructionDescription>
9   </instruction>
```

Listing 11: Instruction description.

This is rendered in Figure 19.

100 FR *Paddle on your forehead*   1

Figure 19: Additional description for an instruction.

Structure

We already encountered the `<repetition>` element on page 24. This element does not specify an instruction, but it defines the program's structure. In the case of a `<repetition>`, it determines that the enclosed elements have to be done multiple times. There are other structural elements.

Segments

Most swim training programs have segments, such as warm up, main set or warm down. These segments can easily be added to a swiML program (see Listing 12).

```
1   <instruction>
2       <segmentName>Warm up</segmentName>
3   </instruction>
4   <instruction>
5       <length>
6           <lengthAsDistance>100</lengthAsDistance>
7       </length>
8       <stroke>
9           <standardStroke>freestyle</standardStroke>
10      </stroke>
11  </instruction>
12  </program>
```

Listing 12: Usage of the segment element.

This is rendered with a visual divider (see Figure 20)

Figure 20: Segments divide the program.

```
                          Warm up
100 FR                                                        1
```

Continue

A typical training program structure is to follow swim instructions without taking a rest. This can be accomplished in swiML with the <continue> element. Similar to an <instruction> element, the <continue> element can have one or many <instruction> elements.

```
1   <instruction>
2       <continue>
3           <instruction>
4               <length>
5                   <lengthAsDistance>100</lengthAsDistance>
6               </length>
7               <stroke>
8                   <standardStroke>freestyle</standardStroke>
9               </stroke>
10          </instruction>
11          <instruction>
12              <length>
13                  <lengthAsDistance>200</lengthAsDistance>
14              </length>
15              <stroke>
16                  <standardStroke>backstroke</standardStroke>
17              </stroke>
18          </instruction>
19      </continue>
20  </instruction>
```

Listing 13: Usage of the continue element.

This is rendered with a vertical visual divider (see Figure 21).

Figure 21: Continue swim instructions.

Python

Creating swim training programs with swiML is easy and yet powerful. But swiML is not a full programming language such as Python. We, therefore, created a Python package[12] that allows you to export swiML programs. This allows you to take advantage of a full programming language. In the next chapter we will show you how to use this, and in the Tools chapter on page 191 you will find a guide to helpful tools. It would go beyond the scope of this book to teach you how to program in Python, but I hope that the examples provided will give you a first idea of the power of swiML when combined with Python.

[12]https://pypi.org/project/swiml_python_xml/

Limitations

The swiML language tries to formalise many aspects of swimming. Similar to many other attempts at formalising and modelling reality, this will never completely succeed. The best model of reality is reality. Still, swiML aspires to cover 80% of swimming practise. There are some challenges that swiML cannot yet handle. For example, swimming strokes and drills are known under different names. We intend to offer support for synonyms in future releases.

It would also be desirable to have support for multiple languages. Localisations are on the development map for swiML. If you would like to contribute to any of the open issues, then please do get in touch.

It has to be acknowledged that swiML already has some built in ambiguity. When swimmers are instructed to swim the stroke "any" or "notFreestyle, then only the swimmers will know what they actually swam. Fitness trackers, such as the Apple Watch or Garmin's watches, are able to record what swimmers actually swam (see Figure 22). While their tracking is imperfect, it does give swimmers good insights into their training performance.

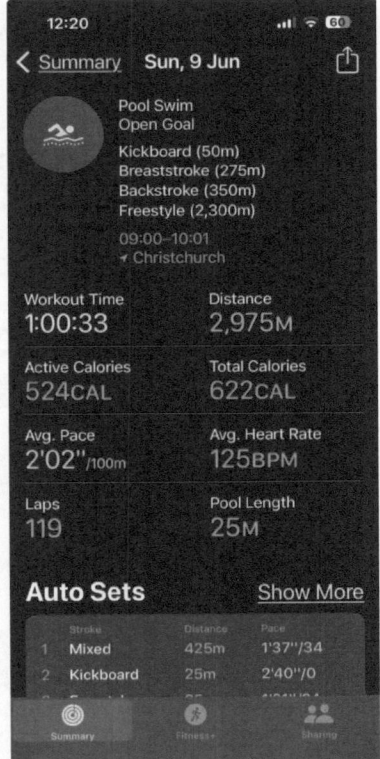

Figure 22: The Apple Watch can track swimming sessions and report on them afterwards.

Conclusions

This chapter gave you a first insight into the use of swiML. Far more details are available in the swiML Schema Reference chapter on page 207. In the next chapter we will put swiML to use by exploring interesting training patterns.

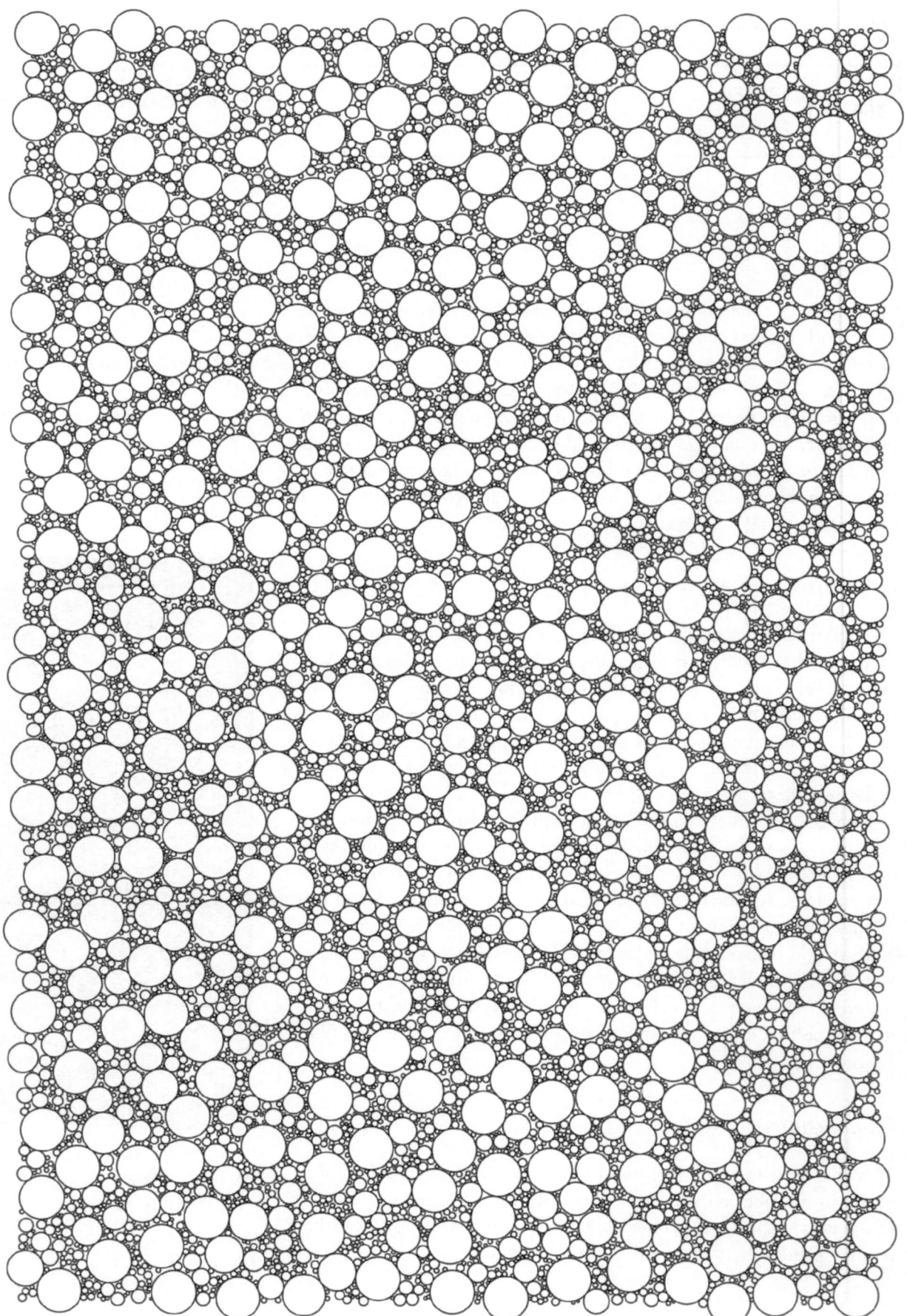

Patterns

Summary

This chapter covers sophisticated patterns in mathematics and computer programming. It includes their historical and cultural context as well as examples of how to formalise them. These patterns are then translated into swim training programs using swiML and the Python programming language.

Swim training is normally broken into patterns. For example, we may swim four times 100 meters freestyle. Patterns that span across a whole training set make it easier for swimmers to remember them. The art of the coach is to create patterns that are, on the one hand, repetitive enough for swimmers to remember while also being interesting to swim. This chapter introduces patterns that are not only suitable for swimming but that will also exercise the mind. The swimmers can think about their underlying ideas while swimming up and down in the pool.

Figure 23: Square tiles in a swimming pool.

Impossible Squares

Modern pools often have 10 lanes, each 2.5 meters wide. This gives the pool a square shape with a width of 25 meters, which results in an area of $25 \times 25 = 625m^2$. When the pool uses square tiles, then interesting patterns emerge. If we ignore the gaps between the tiles and a single tile is 25 cm, then the pool will have $100^2 = 10,000$ tiles. This grid of tiles is ideal for thinking about squares (see Figure 23). The corners at which the tiles connect give us a grid of points (see Figure 24). The rule for drawing is that we can only draw lines by connecting points.

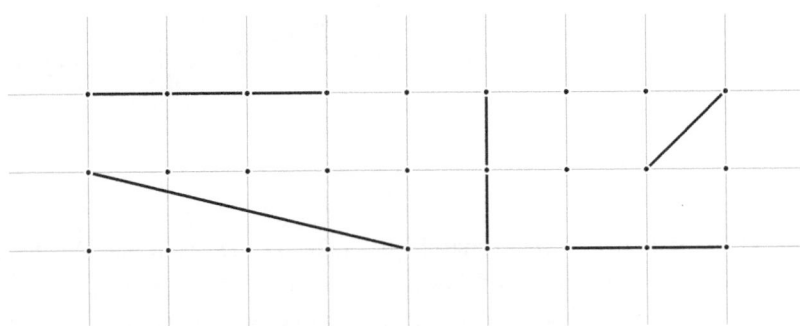

Figure 24: Lines can be drawn onto the grid by connecting points.

If we draw a square by moving one tile across and one tile down then we get a square with the surface of $1 \times 1 = 1$ tiles (see Figure 25i). In our modern pool this would be $25^2 = 625$ square centimetres or 0.0625 m^2. We can also easily draw a 2×2 square with the surface of 4 tiles (see Figure 25ii) or a 3×3 square with a surface of 9 tiles (see Figure 25iii). Any square with the surface a^2 can be easily drawn where a is the number of tiles for the side of the square. Drawing square numbers on a grid of squares is an obvious pattern. Let's call this type of squares "regular squares" and give it the letter R as its symbol.

Figure 25: Drawing squares on a square grid.

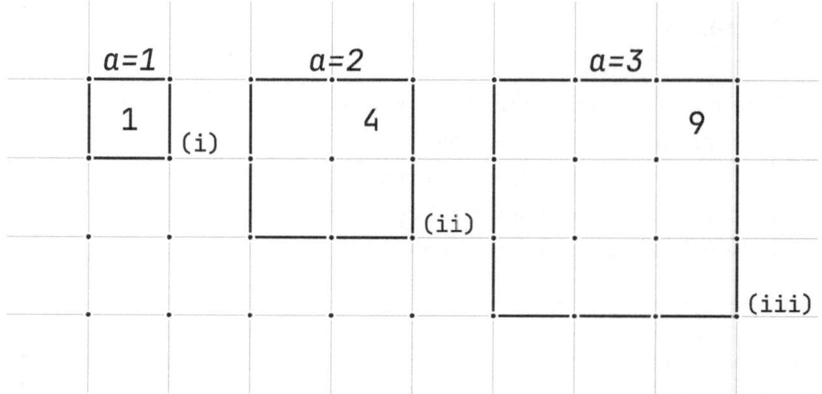

But what about squares that do not have the square number for its surface, such as 2 or 10? They can be drawn by slanted squares (see Figure 26). For a square with the surface of 2 we move one tile across and one tile down (see Figure 26i). We can still intuitively see that a square with the surface of 2 is possible.

Figure 26: Drawing slanted squares on the grid.

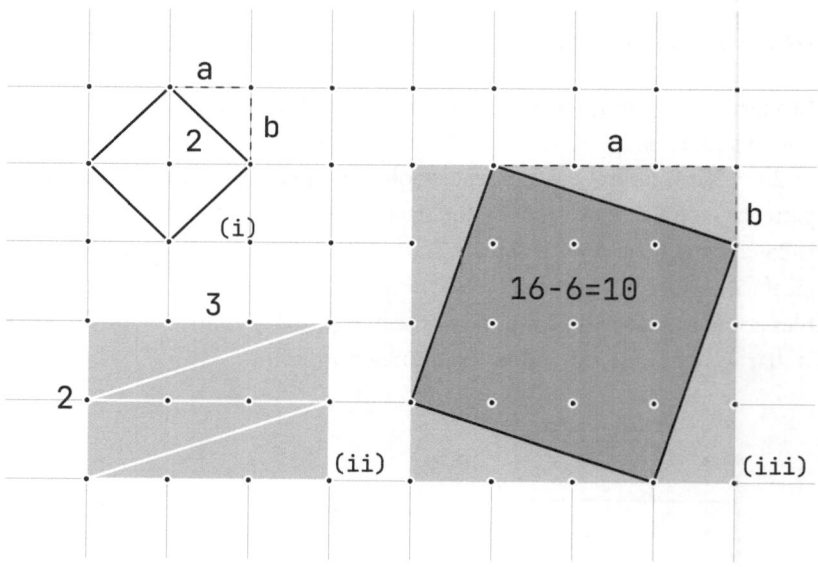

But what about any slanted square? Let's take another example. We can draw a square by moving three tiles across and one down (see Figure 26iii). To easily work out the surface of this square we simply subtract the surface from the four triangles (light grey) from the enclosing square. The enclosing square has a surface of $4 \times 4 = 16$ tiles. The four triangles put together have a surface of $3 \times 2 = 6$ (see Figure 26ii). So a square of three across and one down is $16 - 6 = 10$ tiles. Let's call this type of squares "slanted squares" and give it the letter S as its symbol. There is an even easier way to find the surface for any square. The enclosing square has got a surface of:

$$(a + b)^2$$

The four triangles have the surface of:

$$4 \times \frac{ab}{2}$$

Which can be simplified to:

$$2ab$$

We now have to subtract the surface of the triangles from the surface of the enclosing square:

$$(a + b)^2 - 2ab$$

We can expand this to:

$$(a + b) \times (a + b) - 2ab$$
$$a^2 + ab + b^2 + ab - 2ab \qquad (2)$$
$$a^2 + 2ab + b^2 - 2ab$$

And simplify it to:

$$a^2 \cancel{+ 2ab} + b^2 \cancel{- 2ab}$$
$$a^2 + b^2 \qquad (3)$$

For the "regular squares" $b = 0$, and hence, we are left with the simple square for one of its side a. The slanted squares allow us to draw more squares. So far, we have squares with a surface of $1, 2, 4, 9, 10$. But what about a square with a surface of 3? The short answer is that we cannot draw a square with the surface of three using the grid (see Figure 27). The circle never intersects with a point in the grid.

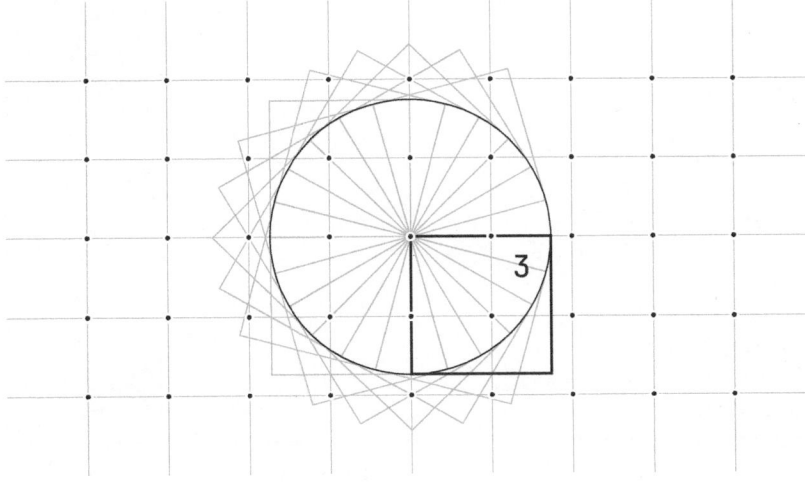

Figure 27: It is impossible to draw a square with a surface of three following the drawing rules. All these squares are illegal.

There are no natural numbers for a and b that result in a surface of 3:

$$a^2 + b^2 = 3$$

Let's try some possibilities starting with the smallest numbers for a and b:

$$0^2 + 1^2 = 1$$

$$1^2 + 1^2 = 2$$

If either a or b is 2 we already get:

$$2^2 + 0^2 = 4$$

By exhaustively trying the possible combinations, we can conclude that a square with the surface of three is impossible. We have found our first impossible square. Let's call this type of square an "impossible square" and give it the letter I as its symbol. Here is a short sequence of these impossible squares: $3, 6, 7, 11, 12, 14, 15, 19, 21, 22, 23, 24, 27...$ To understand which numbers are impossible, we could write a short computer program that would test all the possible combinations for a and b to find the desired surface area of n (see Listing 14).

This sequence is registered with the On-Line Encyclopedia of Integer Sequences as A022544 (https://oeis.org/A022544)

```python
def sum_square(n) :
  # start with a=0 for regular squares
  a = 0
  # loop through all values of a
  # until the square of a is n
  while a * a <= n :
    # start with b=0 for regular squares
    b = 0
    # loop through all values of b
    # until the square of b is n
    while(b * b <= n) :
      # the test if this combination of a and b
      # results in the desired n
      if (a * a + b * b == n) :
        print(a,"^2 +",b,"^2")
        # return back true for having found
        # a combination for a and b that works.
        return True
      # increment b by one for the next loop
      b = b + 1
    # increment a by one for the next loop
    a = a + 1
  # if both loops complete without having found
  # a combination for a and b then the function
  # returns false
  return False
# n is the surface area of the square
n = 18
if (sum_square(n)) :
  print("possible square")
else :
  print( "impossible square")
```

Listing 14: Sum of squares code.

Bonus

The keen observer will have noticed that the calculation for the surfaces of a slanted square is of course a proof for the Pythagorean theorem (see Figure 28).

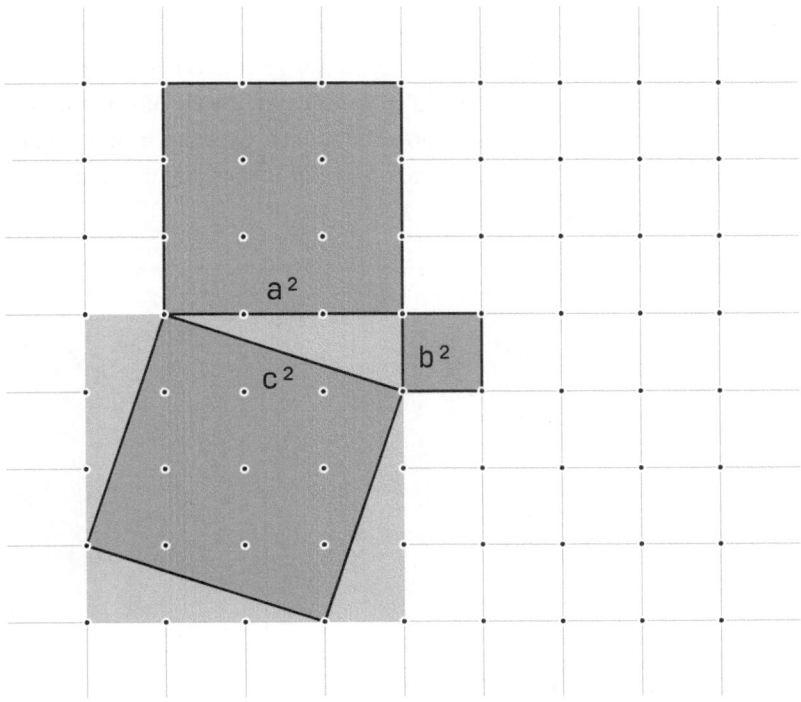

Figure 28: The Pythagorean theorem is the basis for the calculation of the surface of a slanted square.

Figure 29: Pierre de Fermat (6 December 1607 – 12 January 1665) was a French mathematician.

Finding Impossible Squares

This brute force search can quickly become a time consuming exercise since the number of possible combinations for a and b grows quickly. It is desirable to have a quicker test to determine which squares are impossible. The field of number theory has found an answer for this problem, but it would go too far to explain exactly how this is derived. Here, we have to be content with simply accepting this method, which is based on the work of French mathematician Pierre de Fermat (see Figure 29). Fermat's theorem on the sum of two squares states that:

Fermat's theorem 1 *Any positive number n is expressible as a sum of two squares if and only if in the factorisation of n, every factor of the form (4k + 3) occurs an even number of times.*

For this test, the number in question has to be factorised. Let's take the example of 11. Since it is a prime number, it can only be factorised to 1 and 11. The test is if any of the factors are of the form:

$$(4k + 3)$$

where k is a natural number. 11 can be expressed with $k = 2$:

$$(4 \times 2 + 3) = 11$$

The second test is if there is an even or uneven number of this factor present. In this case, we only have one factor, 11×1. Since 1 is an uneven number, this is an impossible square. If we factorise 6 we get 2×3 and 6×1. The factor 3 can be expressed with $k = 0$ as:

$$(4 \times 0 + 3) = 3$$

There is also only one factor three present, which is an uneven number, so this is an impossible square. But what about 9? It also has a factor of 3. It does, however, occur twice, $3 \times 3 = 9$. Since the count of this factor is an even number, this is not an impossible square. 27, on the other hand, has the factors $3 \times 3 \times 3 = 27$. Here, the count is uneven, and hence this is an impossible square.

```python
from collections import Counter

def is_sum_of_two_squares(n):
    # Get the prime factorization of n
    prime_factorization = Counter()
    i = 2
    while i * i <= n:
        while n % i == 0:
            prime_factorization[i] += 1
            n = n // i
        i += 1
    if n > 1:
        prime_factorization[n] += 1

    # Check that every prime of the form (4k + 3) occurs an
    # even number of times
    for prime, count in prime_factorization.items():
        if prime % 4 == 3 and count % 2 != 0:
            return False

    return True

print(is_sum_of_two_squares(14)) # False
print(is_sum_of_two_squares(15)) # True
```

Listing 15: Fermat's theorem test.

Both the brute force and Fermat's theorem method will produce results in an acceptable time for any length that you might be able to swim. Still, the approach using Fermat's theorem will be much faster for surfaces of very large squares. But even the Fermat's theorem approach has considerable limitations. Factorising a non-trivial number remains a difficult task, so difficult that much of cryptography is based upon this kind of activity.

Swimming the Impossible Squares

We can now compose our sequence of regular squares (R), slanted squares (S) and impossible squares (I) (see Equation 4). We can use the surface area as the count of laps to swim. We then swim Backstroke (BA) for every

Regular Square, Breaststroke (BR) for every Slanted Square and Freestyle (FR) for every Impossible Square.:

1	2	3	4	5	6	7	8	9	10	11	12	13	14	15	16	
R	S	I	R	S	I	I	S	R	S	I	I	S	I	I	R	(4)
BA	BR	FR	BA	BR	FR	FR	BR	BA	BR	FR	FR	BR	FR	FR	BA	

We can now write a Python program to generate this sequence at any length we desire. The variable `length_program` in line 83 defines the maximum surface area we want to consider. The `find_squares` function (line 28) is at the heart of determining the status of each surface. To check if the surface is a regular square, it calls the `is_square_number` function (line 23). If it is no, it proceeds to the else-if statement (line 34) that checks Fermat's theorem (see Theorem 1 on 35). It then generates a list that is used in the `create_swiML_instructions` function (line 43). The instructions are then written into the swiML XML file (line 80). We can create programs of any length with this Python program. We only need to change the `length_program` and run it again.

Find this Python program in our repository. You can download the Python program from our repository.

```python
1   from collections import Counter
2   import math, swiML as swiML
3   def is_sum_of_two_squares(n):
4       # Get the prime factorization of n
5       prime_factorization = Counter()
6       i = 2
7       while i * i <= n:
8           while n % i == 0:
9               prime_factorization[i] += 1
10              n = n // i
11          i += 1
12      if n > 1:
13          prime_factorization[n] += 1
14      # Check that every prime of the form (4k + 3) occurs an
15      # even number of times
16      for prime, count in prime_factorization.items():
17          if prime % 4 == 3 and count % 2 != 0:
18              return False
19      return True
20
21  def is_square_number(num):
22      # Check if the square root of the number is an integer
23      square_root = math.sqrt(num)
24      return square_root.is_integer()
25
26  def find_squares(number):
27      impossible_squares=[]
28      i=1
29      # check for every number if it is regular, slanted or impossible
30      # return a list of the result
31      while i <= number:
32          if is_square_number(i):
33              impossible_squares.append(["backstroke","regular"])
34          elif is_sum_of_two_squares(i):
35              impossible_squares.append(["breaststroke","slanted"])
36          else:
```

```
37              impossible_squares.append(["freestyle","impossible"])
38          i+=1
39      return impossible_squares
40
41  def create_swiML_instructions(my_list):
42      my_instruction_list=[]
43      i=1
44      # write an instruction for each list item
45      # return list of instructions
46      while i in range(len(my_list)+1):
47          my_instruction_list.append(swiML.Instruction(
48              length=('lengthAsLaps',i),
49              stroke=('standardStroke',my_list[i-1][0]),
50              instructionDescription=(my_list[i-1][1]),
51              rest=('afterStop','PT0M15S')
52          ))
53          i+=1
54      return my_instruction_list
55
56  def write_program(myInstructions):
57      warmUp=swiML.Instruction(
58          length=('lengthAsDistance',400),
59          stroke=('standardStroke','any'),
60          intensity=('startIntensity',('zone','easy')),
61      )
62      # warm up instructions
63      myInstructions[:0]=[swiML.SegmentName('WarmUp'),warmUp]
64
65      simpleProgram=swiML.Program(
66          title='Impossible Squares',
67          author=[('firstName','Christoph'),('lastName','Bartneck')],
68          programDescription='Swim regular, slanted and impossible
            ↪ squares.',
69          poolLength='25',
70          creationDate='2024-08-22',
71          lengthUnit='meters',
72          hideIntro=True,
73          swiMLVersion='latest',
74          instructions=myInstructions
75      )
76      # write swiML XML to file
77      swiML.writeXML('impossible-squares.xml',simpleProgram)
78
79  # the maximum number of laps which is equal to the number of
    ↪ instructions given.
80  length_program=16
81  # find if the numbers up to length_program are impossible squares
82  squares_list=find_squares(length_program)
83  # create the swiML instructions based on the squares_list
84  instruction_list=create_swiML_instructions(squares_list)
85  # write the swiML XML program to disk
86  write_program(instruction_list)
```

Listing 16: Python program that generates the Impossible Squares training program.

WarmUp		
400 Any Easy		1
1 laps BK ⊙0:15 *regular*		2
2 laps BR ⊙0:15 *slanted*		3
3 laps FR ⊙0:15 *impossible*		4
4 laps BK ⊙0:15 *regular*		5
5 laps BR ⊙0:15 *slanted*		6
6 laps FR ⊙0:15 *impossible*		7
7 laps FR ⊙0:15 *impossible*		8
8 laps BR ⊙0:15 *slanted*		9
9 laps BK ⊙0:15 *regular*		10
10 laps BR ⊙0:15 *slanted*		11
11 laps FR ⊙0:15 *impossible*		12
12 laps FR ⊙0:15 *impossible*		13
13 laps BR ⊙0:15 *slanted*		14
14 laps FR ⊙0:15 *impossible*		15
15 laps FR ⊙0:15 *impossible*		16
16 laps BK ⊙0:15 *regular*		17

Figure 30: The impossible squares training program. You can download this swimming program as a PDF from our repository.

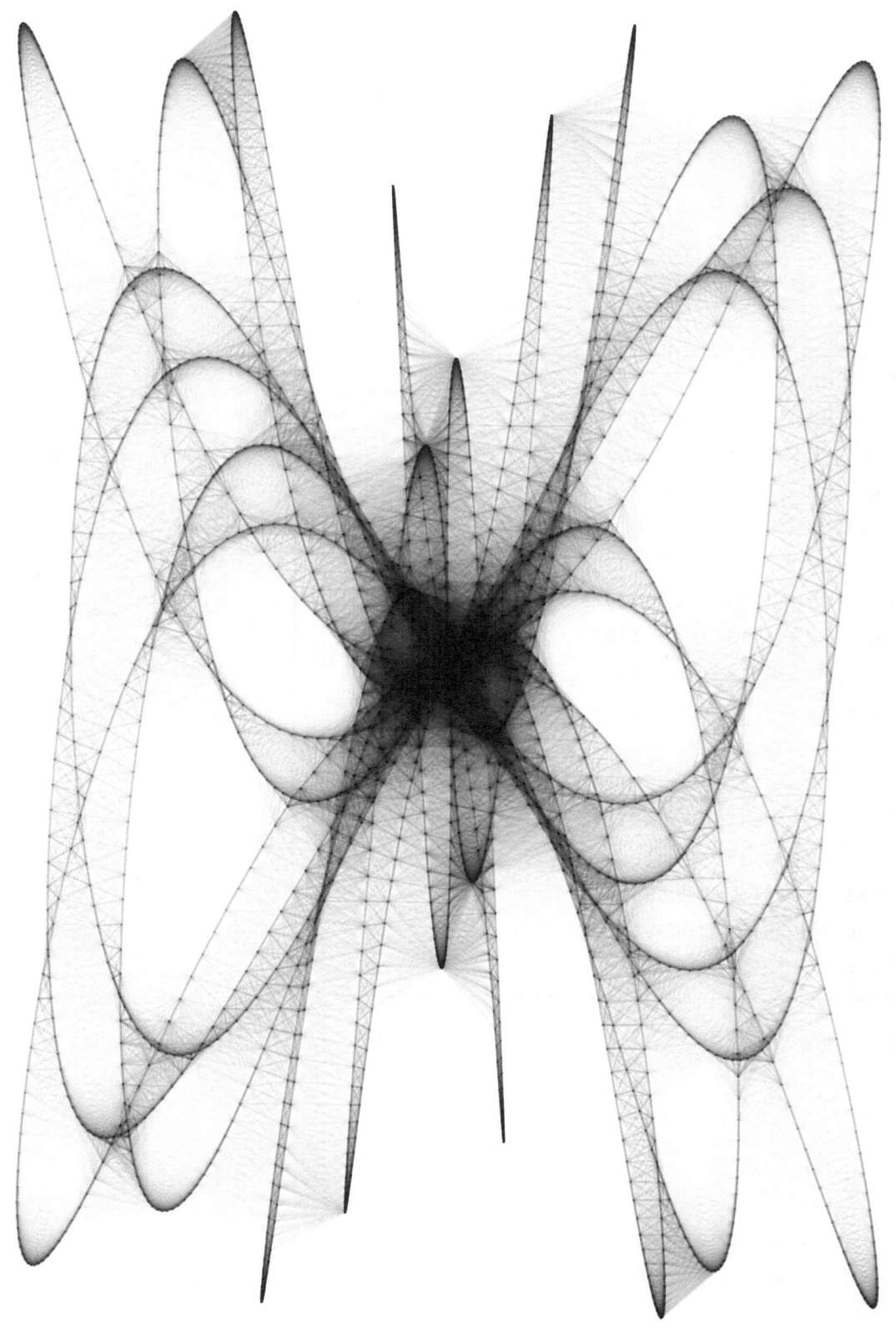

Polyomino

We spend so much time staring at the tiles of the pool that at times I cannot stop imagining shapes and patterns that can be attained by combining tiles. Let's start with a single tile. This is referred to as a monomino (see Figure 31i). This lonely tile has not been combined with any other tile. When you combine two tiles, you get a domino (see Figure 31ii). There is only one type of domino. You could link two tiles on different sides, but they remain identical in shape since you can simply rotate the original shape to get to any other shape. With three tiles, the situation gets slightly more interesting. There are two possible solutions for the trominos (see Figure 31iii). We now also have to exclude mirrored shapes. Solomon Golomb extensively described the Polyomino in 1966 [13] based on earlier work by Henry Ernest Dudeney.

[13]Solomon W. Golomb. *Polyominoes.* Scribner, New York, 1965. ISBN 9780691024448. URL https://search.worldcat.org/title/982644

Figure 31: Polyomino of order 1-3.

$a=1$ $a=2$ $a=3$

(i) (ii) (iii)

Once we combine four tiles we enter familiar territory, assuming that you grew up in the 80s. The five shapes of the tetrominoes are the basis for the popular computer game Tetris (see Figure 32). The five tetrominoes cannot be packed into a rectangle. Only if the pieces are doubled can they be stacked into different sized rectangles, such as a 5×8 rectangle. Otherwise, Tetris would have been too easy. The 12 pentominoes are the basis for the board game Blokus, which is a great way to move puzzling teenagers from computers to a social board game evening.

Figure 32: Tetris is a computer puzzle game created in 1985 by Alexey Pajitnov. It's colourful path to commercialisation has been dramatised in a movie of the same name in 2023.

We can continue to consider the combination of more and more tiles, and their number increases steadily (see Table 1).

Table 1: Name and combinations of the first 12 polyominoes.

n	Names	Combinations
1	monomino	1
2	domino	1
3	tromino	2
4	tetromino	5
5	pentomino	12
6	hexomino	35
7	heptomino	108
8	octomino	369
9	nonomino	1 285
10	decomino	4 655
11	undecomino	17 073
12	dodecomino	63 600

This sequence is registered with the On-Line Encyclopedia of Integer Sequences as A000105 (https://oeis.org/A0001 05)

Swimming Polyomino

We can swim the first four Polyominoes by looking at their geometry shown in Figure 31 and Figure 33.

Figure 33: Polyomino of order 4.

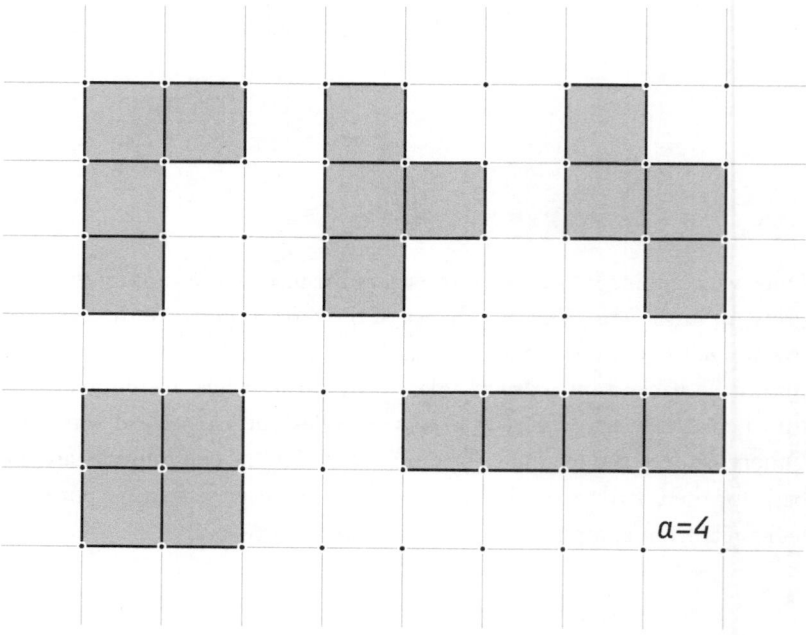

Each tile represents 100 meters. When tiles are horizontally aligned, then the distance multiplies accordingly. The domino is therefore simply 200 meters. The first tromino is also simply 300 meters. The second tromino, however, is split into 200 and 100 meters. Since the 100 meters does not start on the left column, its stroke changes. With this pattern, we can also swim the the five tetromino.

Warm up		
200 Any Easy		1
Monomino		
100 FR ⏱0:15		2
Domino		
200 FR ⏱0:15		3
Tromino		
300 FR ⏱0:15		4
300 as	**200** FR	5
	100 BK	6
Tetromino		
	200 FR	7
400 as	**100** FR	8
	100 FR	9
	100 FR	10
400 as	**200** FR	11
	100 FR	12
	100 FR	13
400 as	**200** FR	14
	100 BK	15
400 as	**200** FR	16
	200 FR	17
400 FR		18
Warm down		
100 Any Easy		19

Figure 34: The Polyomino program. You can download the program from our repository.

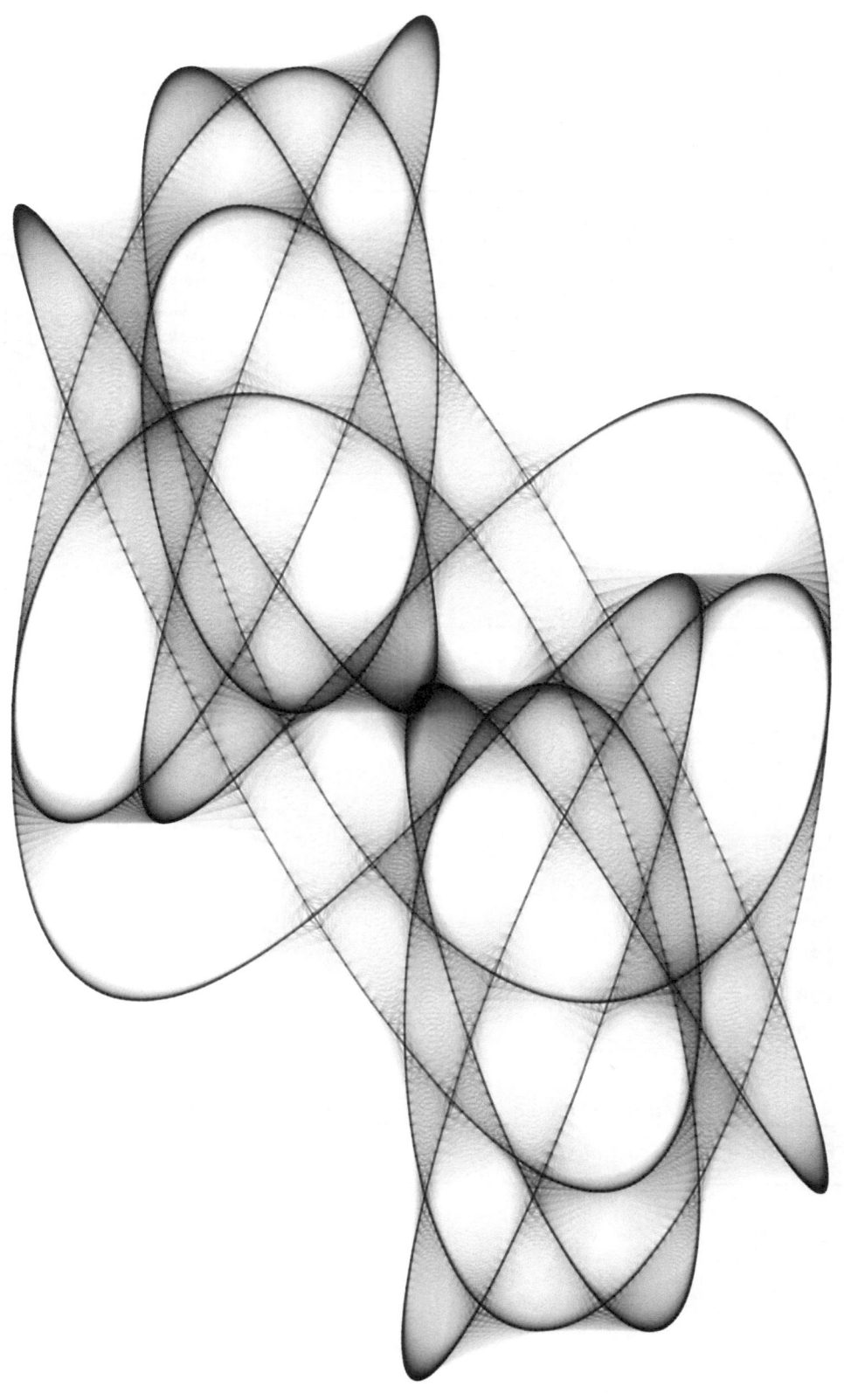

Einstein

While staring at the tiles of a pool for hours and days, you might even stop to wonder why the tiles are shaped rectangularly or even as squares. Too often do we take what is in front of us as optimal. This brings up the question of what we mean by optimal. The established definition for optimal tiling is that it leaves no gaps. But this also holds true for other tile shapes, such as squares and triangles (see Figure 35).

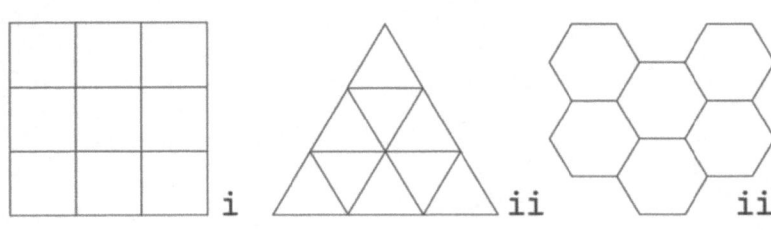

A more differentiating criterion is the relationship of the perimeter to its surface. For a square of side length $a = 1$, the perimeter efficiency E for Area A and Perimeter P given a would be:

$$A = a^2 = 1 \times 1 = 1$$
$$P = 4 \times a = 4 \times 1 = 4$$
$$E = \frac{A}{P} = \frac{1}{4} = 0.25$$

(5)

For an equilateral triangle of side length $a = 1$, this would be:

$$A = \frac{\sqrt{3}}{4} \times a^2 = \frac{\sqrt{3}}{4} \times 1$$
$$P = 3 \times a = 3 \times 1 = 3$$
$$E = \frac{A}{P} = \frac{\frac{\sqrt{3}}{4}}{3} = \frac{\sqrt{3}}{3 \times 4} = 0.143$$

(6)

For a hexagon of side length $a = 1$ this would be:

$$A = \frac{3 \times \sqrt{3}}{2} \times a^2 = \frac{3 \times \sqrt{3}}{2} \times 1$$
$$P = 6 \times a = 6 \times 1 = 6$$
$$E = \frac{A}{P} = \frac{\frac{3 \times \sqrt{3}}{2}}{6}$$
$$= \frac{3 \times \sqrt{3}}{6 \times 2}$$
$$= \frac{\not{3} \times \sqrt{3}}{\not{6} \times 2}$$
$$= \frac{\sqrt{3}}{4} = 0.432$$

(7)

The Hexagons' perimeter efficiency is much higher than that of the other regular tiles. They cover the largest area for the least total perimeter compared to other shapes like squares or triangles. But this is only one definition of optimal.

Another perspective on what an optimal tile is could be how interestingly they are shaped. Triangles and hexagons are already a slight improvement over staring at rectangles, but we can do so much better than this. There are exactly 28 ways to tile a 2D surface with asymmetric tiles [14]. Figure 36 shows one of these 28, the CGCG tile.

[14]Heinrich Heesch and Otto Kienzle. *Flächenschluss; System der Formen lückenlos aneinanderschliessender Flächteile.* Springer, Berlin„ 1963. ISBN 9783642948831. URL https://search.worldcat.org/title/1250086224

Figure 36: Heesch's CGCG tile.

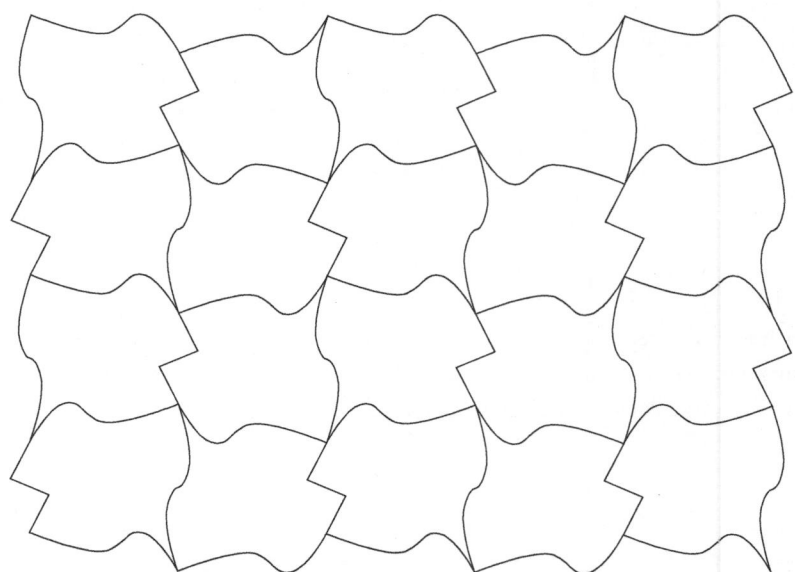

So far, we only considered tiles without any prints on them. This is the most frequent tile in pools. We also only considered a single tile repeating. This does not always work. If you want to use octagons, you will need to combine them with squares (see Figure 37).

Figure 37: Octagon tiles require square tiles to fill the surface.

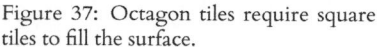
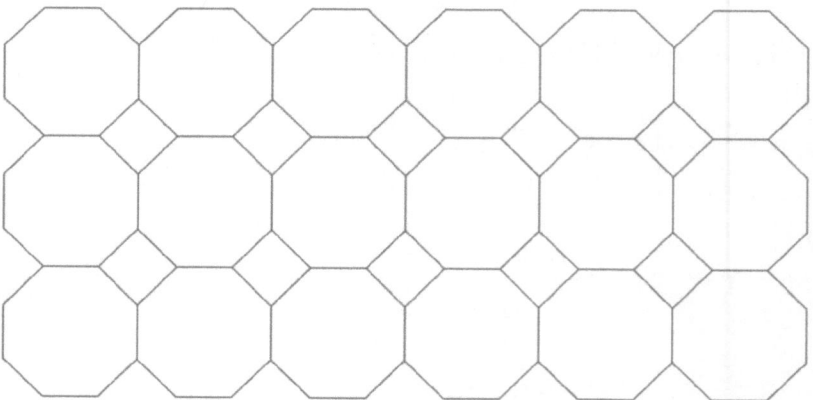

Combining two tiles makes the pattern more interesting, but it still repeats periodically. In 1974, Roger Penrose (see Figure 39) discovered a pair of tiles that does not repeat periodically (see Figure 38). In Mathspeak this type of tiling is called aperiodic.

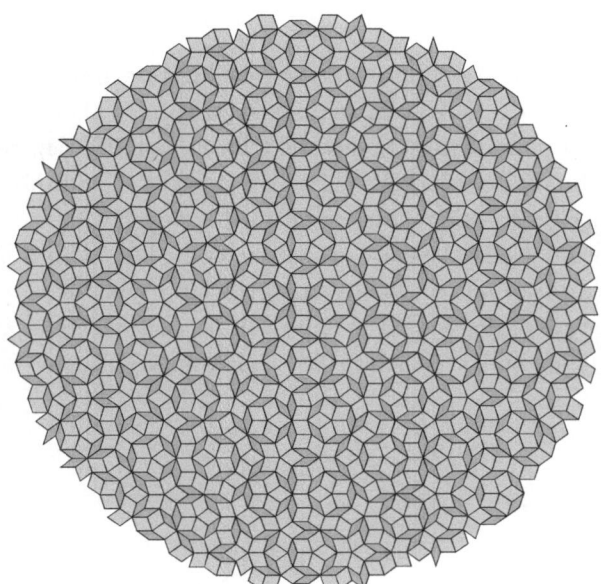

Figure 38: Penrose tilings do not repeat.

But this is not the end of the race for the most interesting yet simple tile. This would be a single aperiodic tile. Mathematicians do have a sense of humour, and hence, this imagined tile was named Einstein. Not after Albert, but after his compound German name structure. Ein (single) Stein (stone). In 2023 David Smith and his team found it! They called their Einstein the "Spectre" (see Figure 40). We now have a single tile that, when assembled, does not produce a repeating pattern. A swimming pool tiled with a Spectre will look uniquely different wherever you look at it. I would argue that this is the optimal tile for keeping a swimmer entertained.

Figure 39: Sir Roger Penrose (∗1931), Nobel Laureate in Physics invented one type of aperiodic tiles. (source: Cirone-Musi, Festival della Scienza)

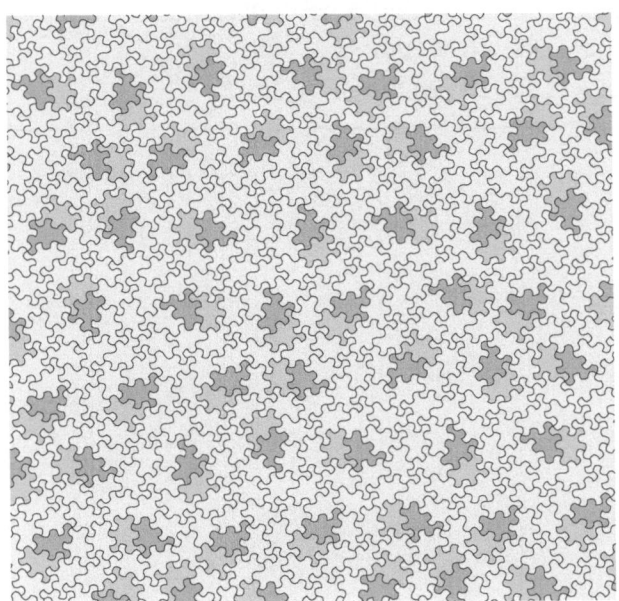

Figure 40: Spectre tilings do not repeat.

Swimming the Einstein

The Spectre might look a bit spooky at first, but at its heart it is just a 14 sided tile. You can easily construct one by starting with a straight line. Copy and rotate it 90 degrees. Copy and rotate it again 120 degrees. These three lines, highlighted in Figure 41i give you a basic unit. You need to reflect and rotate this unit four times and fill the last gap with a straight line to get to the outline of the tile. Replace one of the lines with an arbitrary line (see Figure 41ii). Copy and rotate this arbitrary line while ensuring that it alternates to the inside and outside of the outline (see Figure 41iii) until you end up with the Spectre tile (see Figure 41iv).

Figure 41: Construction of a Spectre tile.

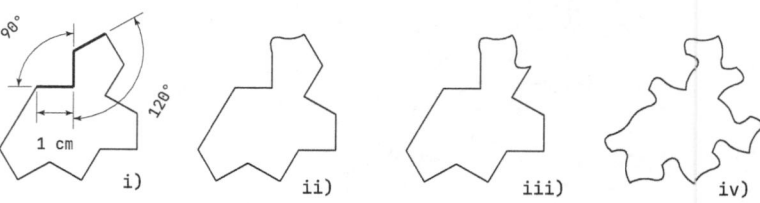

When swimming the Spectre, we simply need to repeat an arbitrary instruction 14 times and rotate it around. We can achieve this by dividing our 2800 meters program into 14 times 200 meters. Half of them are swum in individual medley and the other half in reverse.

Figure 42: The Einstein training program. You can download this swimming program as a PDF from our repository.

Warm up	
200 Any Easy	1
Einstein set	
7 × ⎧ **4 × 50** IM Order @_1:00	2
⎩ **2 × 100** IM Reverse @_1:45	3

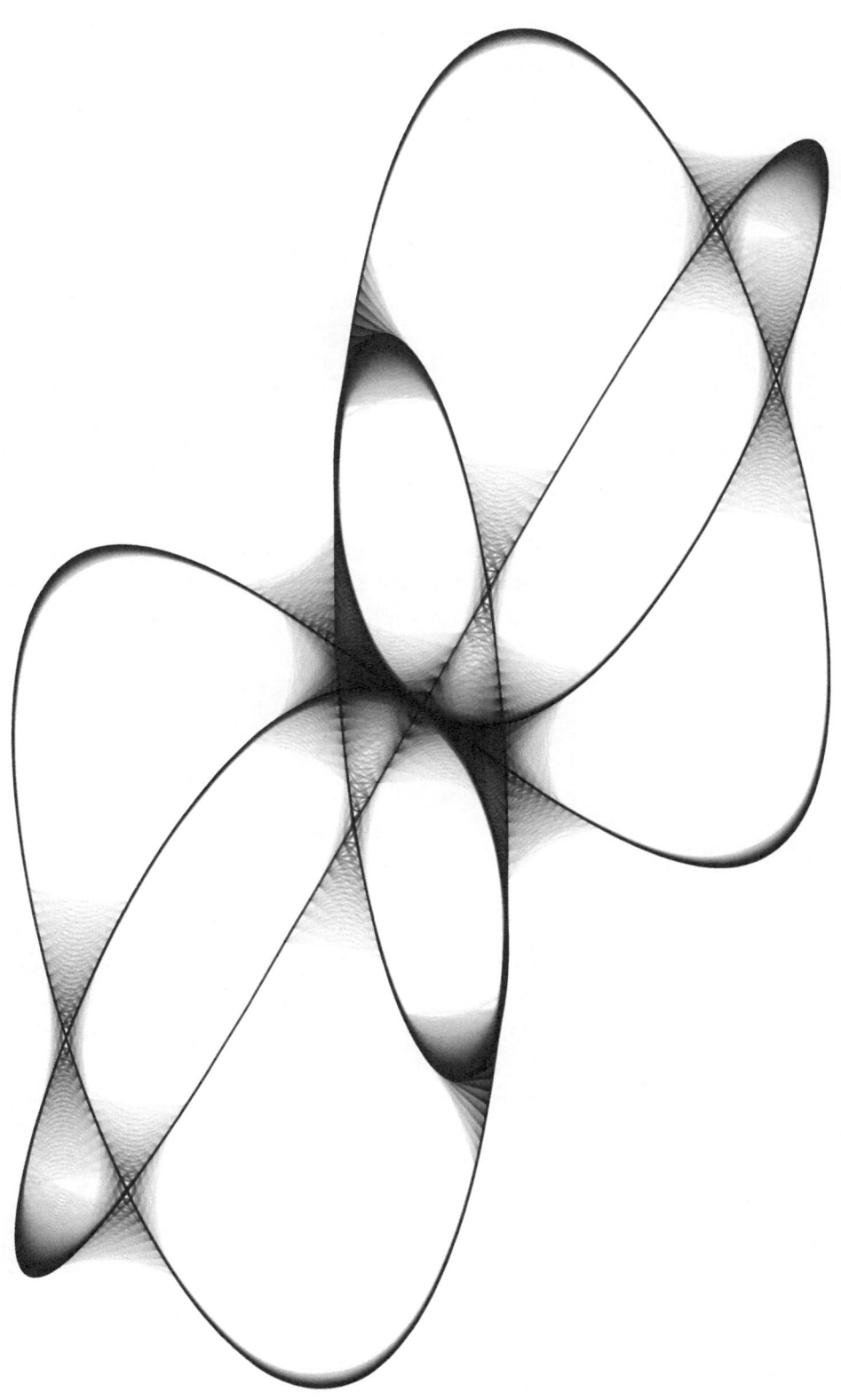

Squares in a Square

The tiles in our square pool can have any size. Instead of asking how many tiles are required to perfectly tile the floor of the pool, we can ask what the smallest square pool is given a number of square tiles.

We can start with the obvious. For any square number of tiles n, the side a of the enclosing square is $\sqrt{n}$. For $n = 1$ square, the enclosing square a is $\sqrt{1} = 1$. The same holds true for $n = 4$ ($a = \sqrt{4} = 2$), $n = 9$ ($a = \sqrt{9} = 3$) and so forth (see Figure 43).

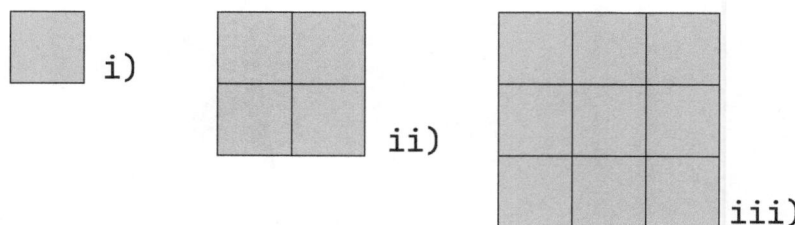

Figure 43: Quadratic number of squares in a square.

The situation becomes slightly less obvious when considering non-quadratic numbers of tiles. For 8, 7 and 6 tiles, there is still no smaller option than $a = 3$ (see Figure 44).

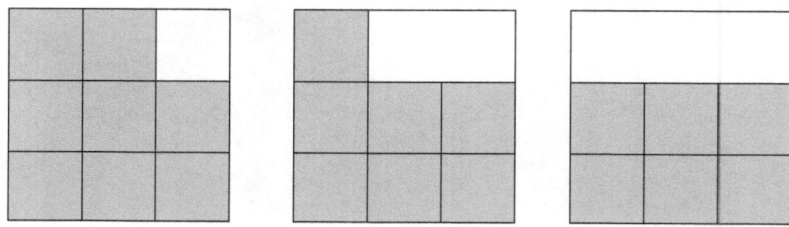

Figure 44: 8,7, and 6 squares in a square.

When considering five tiles, you could be misled to believe that it still only fits into an $a = 3$ square (see Figure 45i). But it turns out that a denser solution is available with $a = 2 + \frac{1}{2} \times \sqrt{2} = 2.707106781$ (see Figure 45ii).

Figure 45: 5 squares in a square.

This solution does still look somewhat elegant, and similar approaches with diagonal tiles exist for any $n = b^2 + b + 3 + \lfloor (b-1) \times \sqrt{2} \rfloor$ tiles which results in $a = b + 1 + \frac{1}{2} \times \sqrt{2}$. Let's consider the case of $b = 2$. This results in a total number of tiles of:

$$n = 2^2 + 2 + 3 + \lfloor (2-1) \times \sqrt{2} \rfloor$$
$$= 4 + 2 + 3 + 1 = 10 \tag{8}$$
$$a = 2 + 1 + \frac{1}{2} \times \sqrt{2} = 3.707106781$$

Figure 46 shows this configuration of 10 tiles with two tiles diagonally in the center.

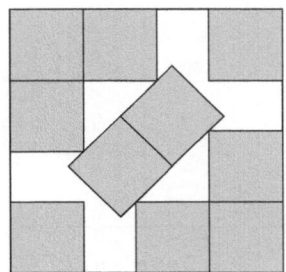

Figure 46: 10 squares in a square.

While this is still somewhat of an elegant solution, the situation becomes far less so when considering 11 tiles (see Figure 47). The squares are no longer angled at 45 degrees, but at 40.182 degrees. There are also tiny gaps between the squares.

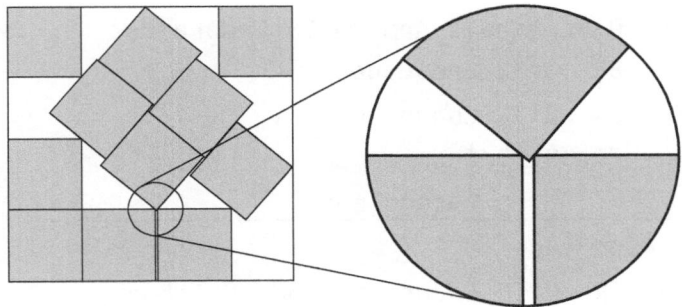

Figure 47: 11 squares in a square.

For some number of tiles, proofs are available for the minimum size of their enclosing squares. Friedman (2012) offered a good overview of the field which is normally referred to as packaging problems. The goal is to find the densest packing of objects in a container. It does not need to be squares. Circles, triangles and many other shapes for tiles and containers have been explored.

Swimming Squares in a Square

We can swim the sequence of squares (see Table 2) in a square by using a as the number of laps. In the non-integer cases, we use a as minutes.

n	1	2	3	4	5	6	7	8	9	10	11	12
a	1	4	4	4	2.7	9	9	9	9	3.7	16	16

Table 2: Sequence of squares in a square.

This results in the following program:

Figure 48: The Squares in a Square training program. You can download this swimming program as a PDF from our repository.

Warm up	
1 laps FL	1
4 laps BK ⏱0:15 Easy	2
4 laps BR ⏱0:15 Easy	3
4 laps FR ⏱0:15 Easy	4
First set	
2:42 FR Race Pace	5
9 laps D FL Single Arm ⏱0:15 Endurance	6
9 laps BK ⏱0:15 Endurance	7
9 laps BR ⏱0:15 Endurance	8
9 laps FR ⏱0:15 Endurance	9
Second set	
3:42 FR Race Pace	10
16 laps FR ⏱0:15 Endurance Pads Pullbuoy	11
16 laps Any Easy	12

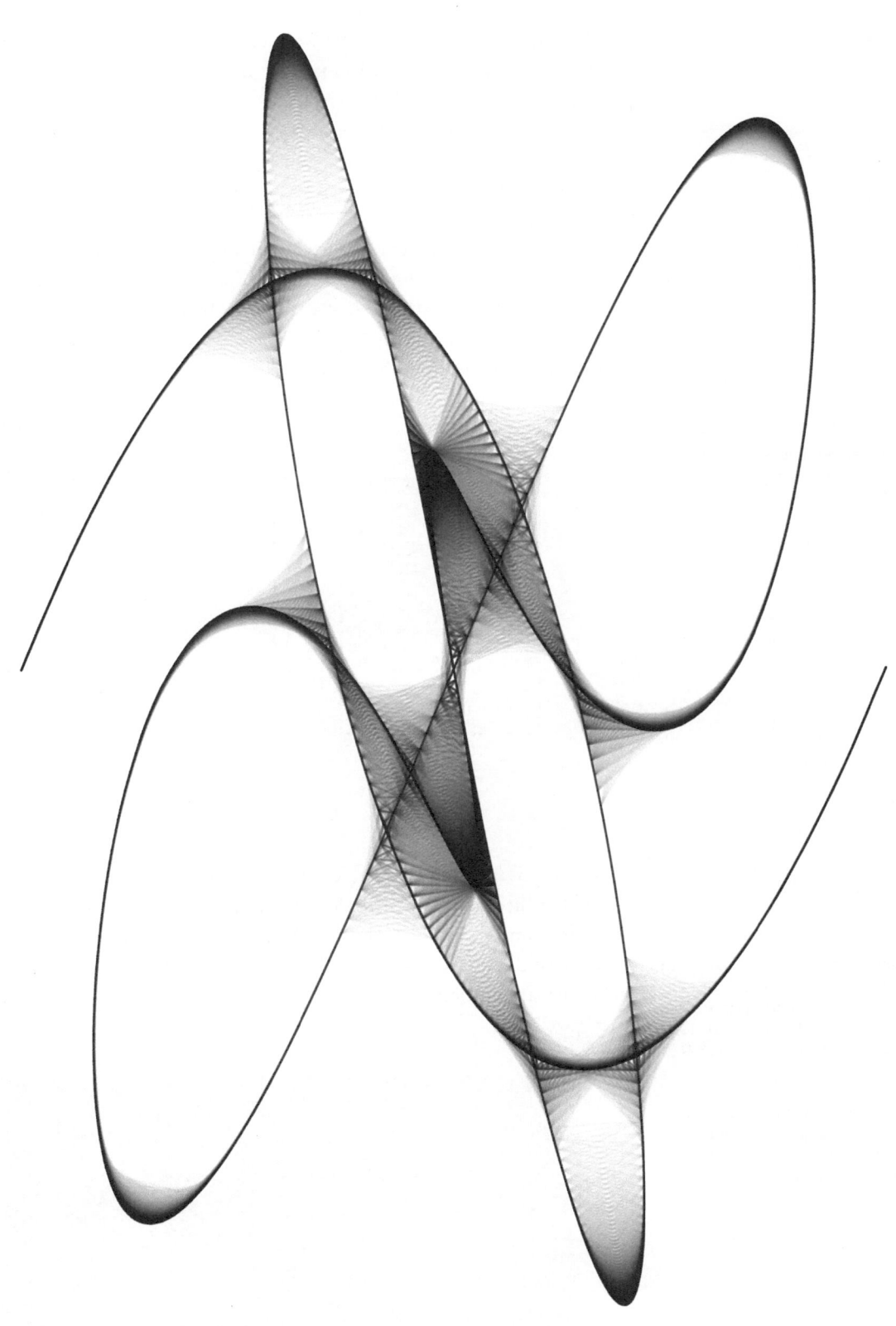

Perfect Squared Square

In our square pool with 100×100 square tiles, we can continue to count tiles until we forget where we were in our training program. Let's focus on a more interesting problem: squaring the square. To be more precise, can we completely fill the surface of the pool with square tiles so that the sizes of all tiles are different? This would obviously make the squared square perfect. If tiles of the same size could occur more than once we would call it an imperfect squared square.

We have to introduce one more constraint. As before, we can only use tiles that can be drawn on the grid used on page 31. Meaning, the tile sizes can only be natural numbers, such as 1, 2, 7, 21, or 42.

Let's consider a smaller 4×4 tiles pool first. We can draw a 3×3 and a 1×1 square into it (see Figure 49). Already drawing the third square would be a repetition of the 1×1 tile.

Figure 49: It is impossible to dissect a 4×4 square into smaller squares.

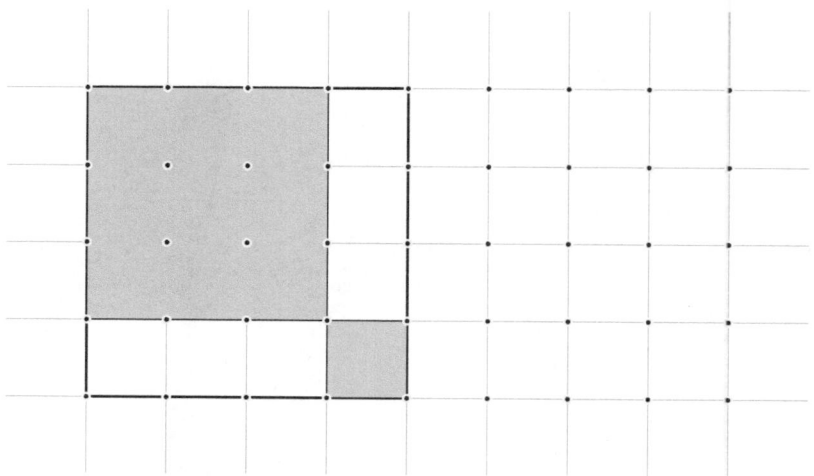

While this looks like a problem that the old Greeks would have already tackled, it only drew attention in the 20th century. Rowland Brooks, Cedric Smith, Arthur Stone and William Tutte published the first Perfect Squared Square in 1940[15]. All four authors of the first published Squared Square are in the 1938 photograph of Trinity Mathematical Society (see Figure 50).

[15] Rowland L Brooks, Cedric AB Smith, Arthur H Stone, and William T Tutte. The dissection of rectangles into squares. *Duke Mathematical Journal*, 7(1):312–340, 1940. DOI: 10.1215/S0012-7094-40-00718-9

Figure 50: The members of the Trinity Mathematical Society in 1938. Notice the complete absence of women (source: Trinity Mathematical Society).

Their Squared Square consisted of 55 unique tiles that took up a pool space of 5468 × 5468. Over the years smaller Squared Squares were found that either used fewer unique tiles or that took up a smaller surface area. The arrival of computers greatly improved the hunt for the smallest Squared Square, and in 1978, Arie Duijvestijn (see Figure 51) found the smallest Perfect Squared Square that uses only 21 unique tiles on a 112 × 112 grid (see Figure 52).

In the same year he also found two Perfect Squared Squares that use 22 unique tiles on a 110 × 110 grid. Since we have proof that this is the smallest possible Perfect Squared Square, we know that it will not fit into our 100 × 100 tiles pool. We need to either expand the pool size to 112 × 112 or we need to subdivide the 25 × 25 meter pool into 112 units, which gives us a grid size of $\frac{2500}{112} = 22.32$

Figure 51: Adrianus Johannes Wilhelmus (Arie) Duijvestijn (10 December 1927 - 21 January 1998) was a Dutch mathematician (source: University of Twente).

Figure 52: The smallest Squared Square consisting of 21 unique squares on a 112 × 112 grid. The numbers denote the length of the side for each square, not its surface.

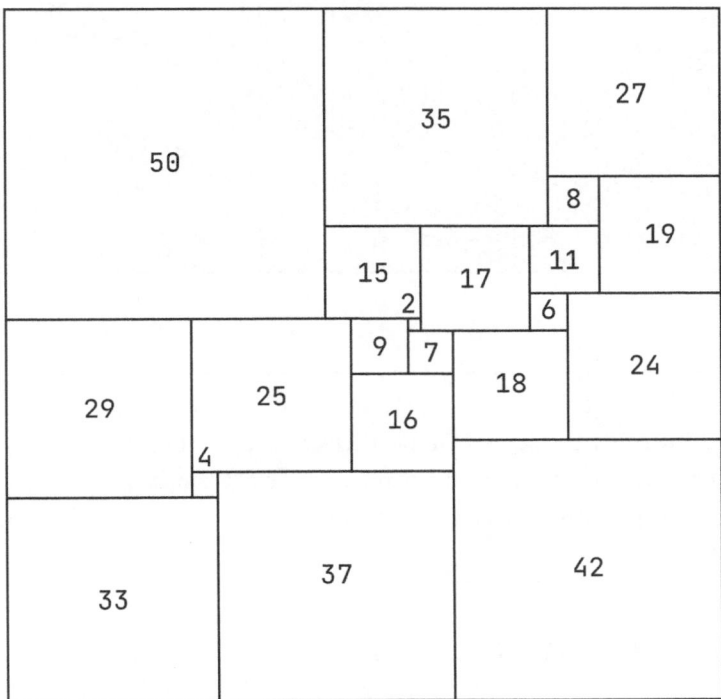

Having to show an image to describe a Squared Square is unsatisfactory. Notice that the caption in Figure 50 names the people in the image starting from the farthest back left moving first to the right. The gentlemen in the photograph were conveniently staged into four rows.

Christoffel Bouwkamp used a similar method to describe Squared, and his nomenclature is often referred to as the Bouwkamp Code. Similar to the gentlemen in the photograph, we start at the top left and first work across. We then move to the next vertical row and start again from left to right (see Figure 53).

The Bouwkamp code for this Squared Square is therefore:

$$(50, 35, 27)(8, 19)(15, 17, 11)(6, 24)(29, 25, 9, 2)(7, 18)(16)(42)(4, 37)(33)$$ (9)

Figure 53: The description of Squared Squares is done by listing the squares from left-to-right and then from top-to-bottom.

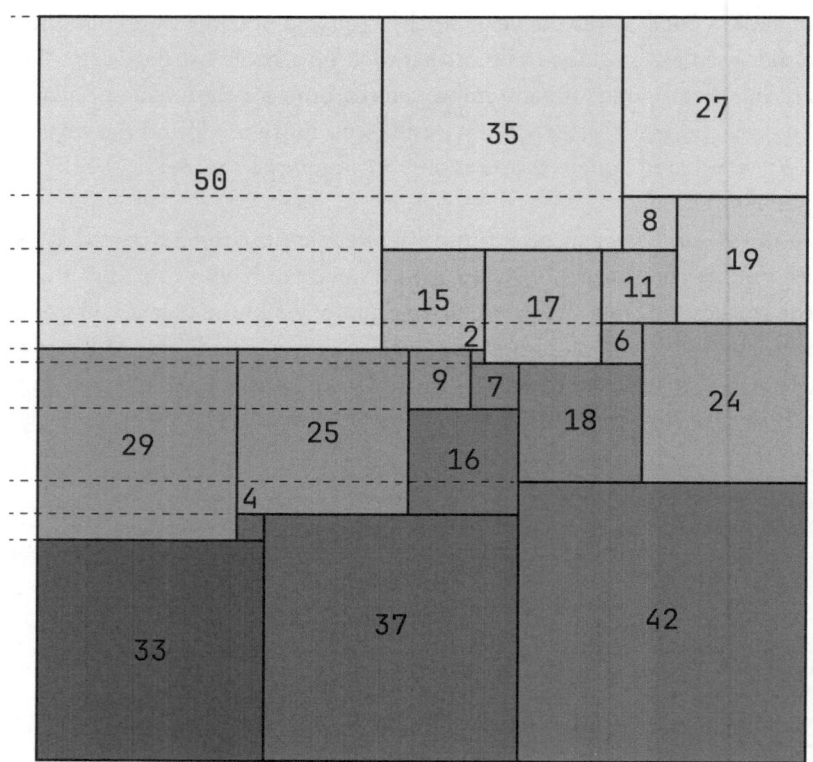

> **Bonus**
>
> So far we only considered Simple Perfect Squared Squares. There is a second type of Perfect Squared Square called "Compound". Squares of this type include a group of tiles that form a square within the square.

Swimming The Perfect Squared Square

We can swim the smallest Perfect Squared Square by using the length of each square as the number of laps (see subsection). Using the Bouwkamp Code, we can associate groups to swimming strokes to give the program more variety. An additional rest between the groups will also be useful.

(50,35,27)	(8,19)	(15,17,11)	(6,24)	(29,25,9,2)	(7,18)	(16)	(42)	(4,37)	(33)
FR	*BK*	*FR*	*BK*	*FR*	*BK*	*IM*	*FR*	*BK*	*FR*

If we sum the squares in Equation 9 we get a total of 434. Swimming this number of laps in a 25 meter pool would add up to 10,850 meters. While not impossible, it is a major commitment. We can aim at a more realistic program by swimming only a third. When we divide the numbers in the sequence by three and round to the nearest integer we get:

(17,12,9)	(3,6)	(5,6,4)	(2,8)	(10,8,3,1)	(2,6)	(5)	(14)	(1,12)	(11)
FR	*BK*	*FR*	*BK*	*FR*	*BK*	*IM*	*FR*	*BK*	*FR*

This results in 145 laps that add up to 3,625 meters in a 25 meter pool.

Unfortunately, this division and rounding does not preserve the uniqueness of the squares. In the reduced Squared Square, the numbers 1, 2, 3, 5, 6, 9 and 12 occur twice.

WarmUp	
400 Any Easy	1
First row	
17 laps FR ⏱0:15	2
12 laps FR ⏱0:15	3
9 laps FR ⏱0:15	4
Second row	
3 laps BK ⏱0:15	5
6 laps BK ⏱0:15	6
Third row	
5 laps FR ⏱0:15	7
6 laps FR ⏱0:15	8
4 laps FR ⏱0:15	9
Fourth row	
2 laps BK ⏱0:15	10
8 laps BK ⏱0:15	11
Fifth row	
10 laps BK ⏱0:15	12
8 laps BK ⏱0:15	13
3 laps BK ⏱0:15	14
1 laps BK ⏱0:15	15
Sixth row	
2 laps BK ⏱0:15	16
6 laps BK ⏱0:15	17
Seventh row	
5 laps IM ⏱0:15	18
Eighth row	
14 laps FR ⏱0:15	19
Ninth row	
1 laps BK ⏱0:15	20
12 laps BK ⏱0:15	21
Tenth row	
11 laps FR ⏱0:15	22

Figure 54: The Squared Square training program. You can download this swimming program as a PDF from our repository.

What Number is Next?

In mathematics, sequences of integers are a popular topic. Often, they are riddles where you are being asked to guess what the next number might be. So let's give this a try:

$$1, 4, 9, 16, 25, \ldots \tag{10}$$

What is the next number? It is, of course, 36 since the sequence is based on the power of two $1^2, 2^2, 3^2, 4^2, 5^2, 6^2$. Let's try another sequence:

$$1, 11, 21, 1211, 111221, \ldots \tag{11}$$

What is the next number? I will give you a few moments to think about it.

[16]J. H. Conway. The weird and wonderful chemistry of audioactive decay. In Thomas M. Cover and B. Gopinath, editors, *Open Problems in Communication and Computation*, pages 173–188. Springer New York, New York, NY, 1987. ISBN 978-1-4612-4808-8. DOI: 10.1007/978-1-4612-4808-8_53

It is 312211. This may or may not come as a surprise to you. If you are surprised, then you are not alone. This sequence catches many off guard. The name of this sequence is "look and say" and was first described in 1987[16]. I will give you few more moments to try it.

This sequence is registered with the On-Line Encyclopedia of Integer Sequences as A005150 (https://oeis.org/A005150)

If you have not yet solved this little riddle, then let's make it explicit.

- 1 is read as off as "one 1", or 11

- 11 is read off as "two 1s" or 21

- 21 is read off as "one 2, one 1" or 1211

- 1211 is read off as "one 1, one 2, two 1s" or 111221

- 111221 is read off as "three 1s, two 2s, one 1" or 312211

The numbers in this sequence grow quickly. Here are the first 12 terms (see Table 3):

Table 3: The first 12 terms of the Look and Say sequence. You can download a Python program to find the terms of the Look and Say sequence from our repository.

Term	Number
1	1
2	11
3	21
4	1 211
5	111 221
6	312 211
7	13 112 221
8	1 113 213 211
9	31 131 211 131 221
10	13 211 311 123 113 112 211
11	11 131 221 133 112 132 113 212 221
12	3 113 112 221 232 112 111 312 211 312 113 211

Swimming Look and Say

The Look and Say sequence has a couple of interesting properties. If we start the sequence with the seed 1, then we will only ever get the digits 1, 2 and 3. The sequence growths in length at approximately 30% per iteration. The 10th element, 13 211 311 123 113 112 211, already has 20 digits. Notice that in this program we use the strokes Nr 1, Nr 2 and Nr 3. This refers to your favourite strokes (see 215).

WarmUp	
400 Any Easy	1
1 × **100** as **1** Nr 1 ⟳0:15	2
2 × **100** as ⎰ **1** Nr 1 ⟳0:15	3
⎱ **1** Nr 1 ⟳0:15	4
3 × **100** as ⎰ **2** Nr 2 ⟳0:15	5
⎱ **1** Nr 1 ⟳0:15	6
1 Nr 1 ⟳0:15	7
5 × **100** as **2** Nr 2 ⟳0:15	8
1 Nr 1 ⟳0:15	9
1 Nr 1 ⟳0:15	10
1 Nr 1 ⟳0:15	11
1 Nr 1 ⟳0:15	12
1 Nr 1 ⟳0:15	13
8 × **100** as **2** Nr 2 ⟳0:15	14
2 Nr 2 ⟳0:15	15
1 Nr 1 ⟳0:15	16
3 Nr 3 ⟳0:15	17
1 Nr 1 ⟳0:15	18
2 Nr 2 ⟳0:15	19
10 × **100** as **2** Nr 2 ⟳0:15	20
1 Nr 1 ⟳0:15	21
1 Nr 1 ⟳0:15	22

Figure 55: The Look and Say training program. You can download this swimming program as a PDF from our repository.

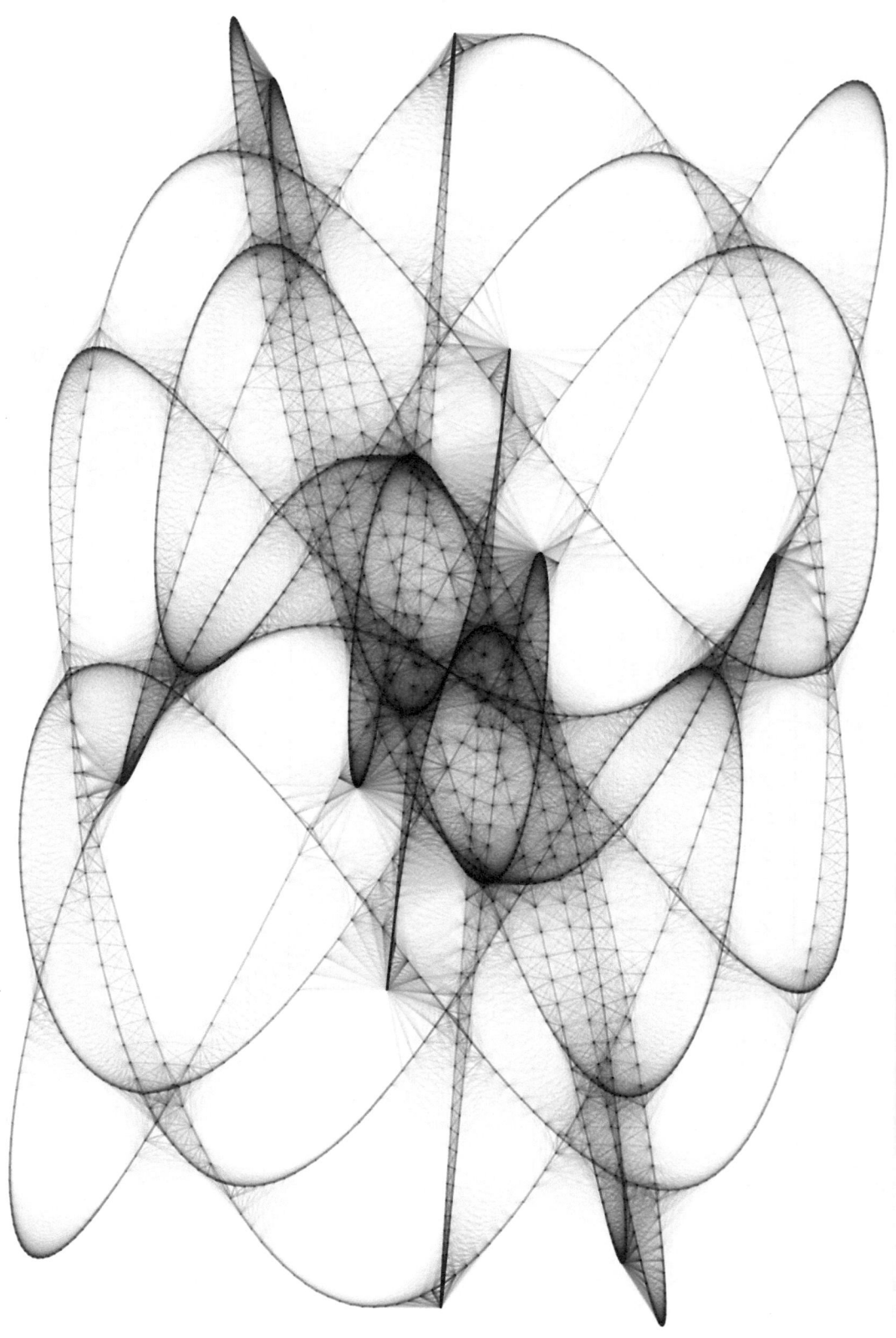

Binary

The decimal system is the most popular numeral system for humans, although it also contains certain irregularities. Instead of saying oneteen we say eleven and instead of twoteen we say twelve. The French have an even more complicated counting system. Their vigesimal numeral system uses base-20. While speaking the numbers might be complicated, they still remain a base 10 system. Meaning that we have 10 symbols ranging from 0 to 9.

Computers use the binary numeral system which is based on base 2. It only knows two symbols: 0 and 1. Counting in binary follows the same pattern as counting in decimal. We first go through all the symbols (0...9) before adding a digit (10). In binary, this works as:

Decimal	Binary
0	0
1	1
2	10
3	11
4	100
5	101
6	110
7	111
8	1000
9	1001
10	1010

(12)

Bonus

The binary system is not the only system used in computers. The hexadecimal system, which uses base 16, is also in use. Latin based languages have only numeral systems ranging from 0 to 9. The hexadecimal system does require six more symbols, and it resorts to characters A to F. Counting in hexadecimals works like this:

Decimal	Hexadecimal	Binary
0	0	00000
1	1	00001
2	2	00010
3	3	00011
4	4	00100
5	5	00101
6	6	00110
7	7	00111
8	8	01000
9	9	01001
10	A	01010
11	B	01011
12	C	01100
13	D	01101
14	E	01110
15	F	01111
16	10	10000
17	11	10001

Here is a Python program to create variations of the Binary Program. You can download the Python program from our repository.

Swimming Binary

One bit of memory in a computer can hold either a zero or a one. With three bits, we can count from 000 to 111. In decimals, this would count from 0 to 7. If you prefer to start counting from 1, then this would be equivalent to counting from 1 to 8. Swimming three bits would then be equivalent to swimming eight times a multiple of three. If we assume this to be 100 meters, then the program would be 8 times 300 meters. We can allocate the number 1 to swimming freestyle and 0 to non-freestyle.

Warm Up			
400 Any Easy			1

Binary set			
		100 Not FR	2
	1 as	**100** Not FR	3
		100 Not FR	4
		100 Not FR	5
	1 as	**100** Not FR	6
		100 FR	7
		100 Not FR	8
	1 as	**100** FR	9
		100 Not FR	10
		100 Not FR	11
8 × **300** as	**1** as	**100** FR	12
		100 FR	13
		100 FR	14
	1 as	**100** Not FR	15
		100 Not FR	16
		100 FR	17
	1 as	**100** Not FR	18
		100 FR	19
		100 FR	20
	1 as	**100** FR	21
		100 Not FR	22
		100 FR	23
	1 as	**100** FR	24
		100 FR	25

Warm Down			
200 Any Easy			26

Figure 56: Swim the Binary program for the number 8. You can download this swimming program as a PDF from our repository.

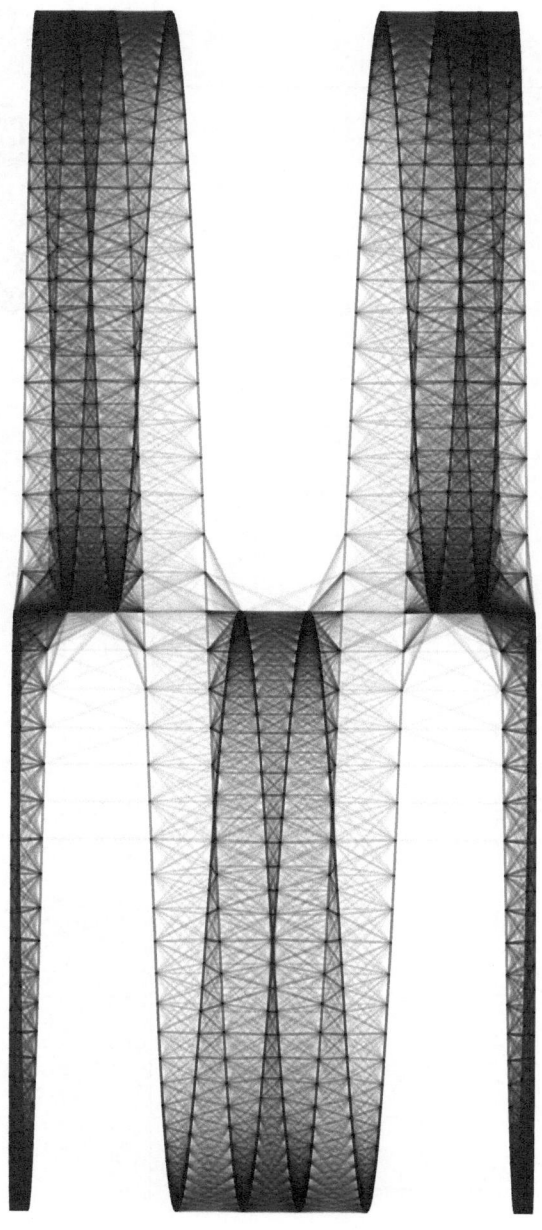

Super Hero Triangles

There are different types of special triangles, such as equilateral, isosceles or right triangles. You might still be familiar with them from your geometry classes. An equilateral triangle, for example, has sides that are all equally long (see Figure 57, left). The Heronian triangles, however, are special amongst the special triangles. The length of each side can be expressed as an integer, and the area is also an integer value (see Figure 57, right)[17].

[17]The Heronian triangles are named after Heron of Alexandria, a Greek mathematician living around 60AD in Alexandria.
Figure 57: The equilateral triangle on the left has integer sides, but an area of 43.3. The Heronian triangle on the right has integer sides and an integer surface area of $A = 84$.

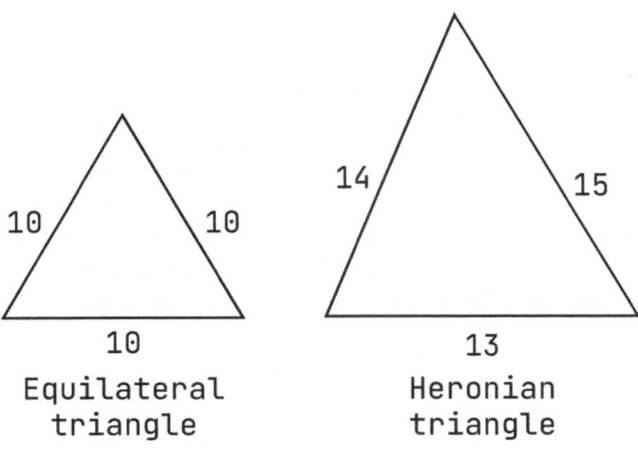

But there are even more special triangles, the Super Hero Triangles. A Super Hero Triangle is a Heronian triangle for which its perimeter is equal to its area[18]. Amongst the indefinite number of triangles, there are exactly five Super Hero Triangles (see Figure 58).

[18]Lubomir Markov. Heronian triangles whose areas are integer multiples of their perimeters. *Forum Geometriorum*, 7:129–135, 2007. URL https://forumgeom.fau.edu/FG2007volume7/FG200718.pdf

Figure 58: The five Superhero Triangles.

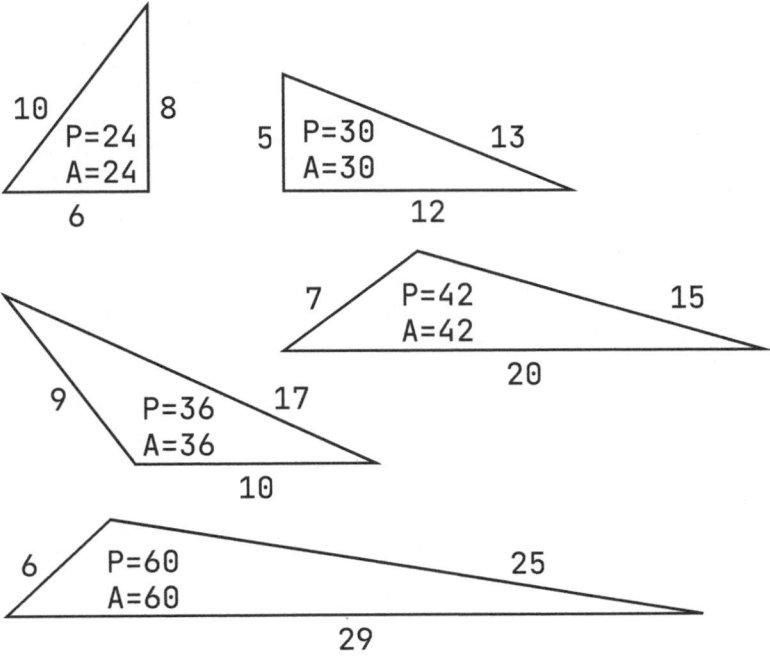

Swimming the Super Hero Triangles

We can swim all three sides of the five triangles. We simply use the integer for each side of each triangle as the number of laps. Table 4 shows the length of each triangle side, the area A and the perimeter P.

Table 4: The length of the triangle sides (a, b, c), area A and perimeter P

Triangle Nr.	a	b	c	$A = P$
1	6	8	10	24
2	5	12	13	30
3	7	15	20	42
4	9	10	17	36
5	6	25	29	60
stroke	nr3	nr2	nr1	

In a 25 meter pool this would result in a 4 800 meter program that could be swum in 90 minutes (see Figure 59). For less ambitious swimmers, each number of laps could be cut in half and then rounded down.

Warm up	
400 Any	1
First triangle	
6 laps Nr 3 ⟳0:15	2
8 laps Nr 2 ⟳0:15	3
10 laps Nr 1 ⟳0:15	4
Second triangle	
5 laps Nr 3 ⟳0:15	5
12 laps Nr 2 ⟳0:15	6
13 laps Nr 1 ⟳0:15	7
Third triangle	
7 laps Nr 3 ⟳0:15	8
15 laps Nr 2 ⟳0:15	9
20 laps Nr 1 ⟳0:15	10
Fourth triangle	
9 laps Nr 3 ⟳0:15	11
10 laps Nr 2 ⟳0:15	12
17 laps Nr 1 ⟳0:15	13
Fifth triangle	
6 laps Nr 3 ⟳0:15	14
25 laps Nr 2 ⟳0:15	15
29 laps Nr 1 ⟳0:15	16

Figure 59: The Superhero Triangles program. You can download this swimming program as a PDF from our repository.

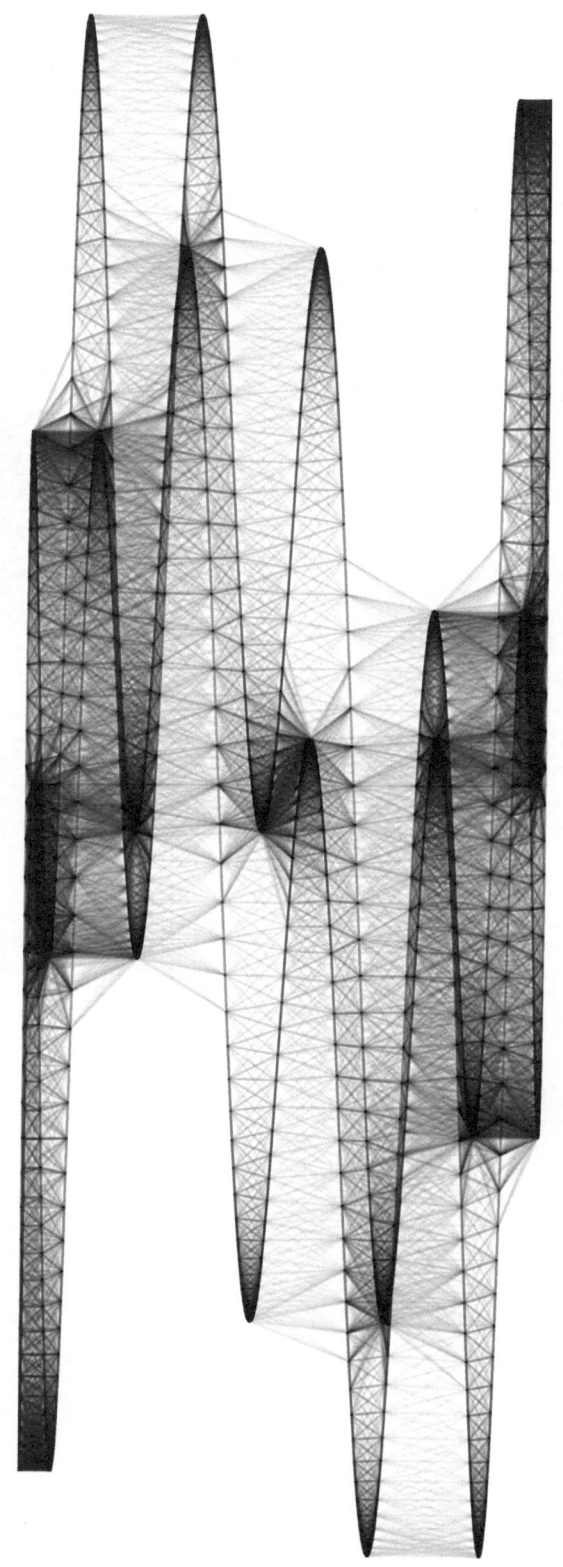

Fibonacci Sequence

The Fibonacci sequence[19] is of striking simplicity, yet it entails a considerable complexity. In Europe, the Italian mathematician Fibbonaci (see Figure 60) first described the sequence in his book Liber Abaci[20].

Here is the start of the sequence:

$$0, 1, 1, 2, 3, 5, 8, 13, 21, 34, \dots \tag{13}$$

Can you guess which number will come next? I will give you a few moments to think about it.

[19]The Fibonacci Sequence is registered with the On-Line Encyclopedia of Integer Sequences as A000045 https://oeis.org/A000045

[20]Laurence Sigler. *Fibonacci's Liber Abaci: A Translation into Modern English of Leonardo Pisano's Book of Calculation.* Springer Science & Business Media, 2002. ISBN 9780387954196. URL https://www.worldcat.org/title/48557588

Figure 60: Fibonacci (circa 1170 – 1250) was an Italian mathematician.

It is 55. The sum of the previous two numbers. It can be expressed with a recursive formula. The first two numbers, f_0 and f_1, in the sequence are set as:

$$f_0 = f_1 = 1 \tag{14}$$

The following numbers are calculated as:

$$f_n = f_{n-1} + f_{n-2} \text{ for } n \geq 2 \tag{15}$$

We already talked about recursion on page 26 when a function is defined by referring to itself. An integer sequence alone does not yet justify attention. There are thousands of possible sequences. There are two possible reasons for its popularity. First, this sequence can be observed in nature. The shell of the Nautilus, for example, seems to resemble that of a spiral generated from the Fibonacci sequence. We can draw such a Fibonacci spiral by first drawing squares of the sequence. We then draw quarter circles in them (see Figure 61).

The spiral that emerges has some resemblance to that of sea shells, in particular that of the Nautilus shell, but Falbo (2005) showed that this is unfortunately not the case. Still, Fibonacci numbers can be observed in the number of spirals in a sunflower or the number of petals on a plant. The lily and iris have three petals while buttercups and wild roses have five. Sunflowers typically have 34 or 55 spirals. We will come to them shortly [21].

[21] Charles Safran. The Fibonacci Numbers. *Chance*, 5(1-2):43–46, 1992. DOI: 10.1080/09332480.1992.11882462

Figure 61: The Fibonacci spiral.

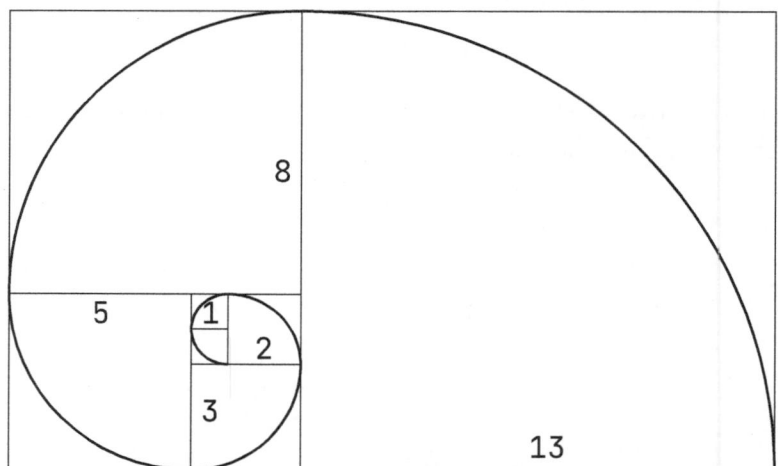

Golden Ratio

The second reason for the popularity of the Fibonacci number is its relationship to the Golden Ratio. The Golden Ratio is often denoted with the Greek letter phi ϕ (pronounced /faɪ/). It is defined as a is to $a + b$ as a is to b (see Figure 62).

Figure 62: The Golden Ratio.

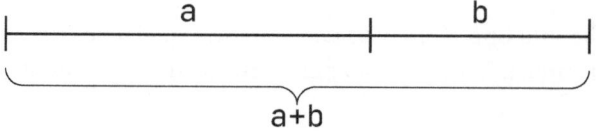

The Golden Ratio can then be expressed for all $a > b > 0$:

$$\frac{a + b}{a} = \frac{a}{b} = \phi \tag{16}$$

To find ϕ we start with the left side of the equation which can be simplified to:

$$\frac{a + b}{a} = \frac{a}{a} + \frac{b}{a} = 1 + \frac{b}{a} \tag{17}$$

$\frac{b}{a}$ is the inverse of $\frac{a}{b}$ which is equal to ϕ as defined in Equation 16. We can therefore substitute $\frac{b}{a}$ with $\frac{1}{\phi}$ leading to:

$$1 + \frac{1}{\phi} = \phi \tag{18}$$

Multiplying both sides of the equation with ϕ brings us to:

$$\phi + 1 = \phi^2 \tag{19}$$

which can be rearranged to:

$$\phi^2 - \phi - 1 = 0 \tag{20}$$

We now have a normal quadratic formula which has the standard form of:

$$ax^2 + bx + c = 0 \tag{21}$$

In Equation 20 we see that $x = \phi, a = 1, b = -1, c = -1$. Quadratic equations can be solved with the formula below. It will go beyond the scope of this book to explain how this formula is derived. For now, let's just accept that this is the accurate solution.

$$x = \frac{-b \pm \sqrt{b^2 - 4ac}}{2a} \tag{22}$$

When we use the values for x, a, b and c we get:

$$\phi = \frac{-1^2 \pm \sqrt{-1^2 - (4 \times 1 \times -1)}}{2 \times 1}$$
$$= \frac{1 \pm \sqrt{1 + 4}}{2} \tag{23}$$

The positive solution simplifies to:

$$\phi = \frac{1 + \sqrt{5}}{2} = 1{,}618\,033\,988\,7... \tag{24}$$

ϕ is an irrational number, which means that it cannot be expressed as a simple fraction. A rational number, such as 2 or 0.5, can be expressed as a fraction, $\frac{2}{1}$ and $\frac{1}{2}$ respectively. But also 0.33333... is a rational number since it can be expressed as $\frac{1}{3}$. An irrational number is a number that is not rational.

You may argue that we just did express ϕ in Equation 24 as a simple equation. But we could only achieve this by using $\sqrt{5}$. $\sqrt{5}$ itself is an irrational number, which means that we have no hope of ever expressing ϕ as a rational number. As a result, the digits of ϕ will continue forever.

The best hope we have is to approximate ϕ. This can be done with the Fibonacci sequence. The ratio of $\frac{f_{n+1}}{f_n}$ is approximating ϕ as n increases (see Equation 25).

f_{n+1}	f_n	$\frac{f_{n+1}}{f_n} \approx \phi$	
3	2	1.500	
5	3	1.666	
8	5	1.600	(25)
13	8	1.625	
$\vdots$	$\vdots$	$\vdots$	
233	144	1.618	

While the Fibonacci Sequence approximates the Golden Ratio, it is not the only sequence of numbers that does so. We can select any two random numbers and start constructing a sequence in the same way as the Fibonacci Sequence. As swimmers, we could choose the two smallest distances we can swim in a typical pool. Table 5 shows how this sequence unfolds and how it approximates the Golden Ratio of 1,618 033 988 7...

Table 5: The Swimming Numbers approximate the Golden Ratio.

Swimming Numbers	Ratio
25	
50	2.000000
75	1.500000
125	1.666667
200	1.600000
325	1.625000
525	1.615385
850	1.619048
1 375	1.617647
2 225	1.618182

The Golden Ratio in Art and Nature

The Golden Ratio is often used in the arts where geometrical arrangements, such as a Golden Rectangle (see Figure 63), that use this ratio are considered to be more harmonious.

Figure 63: The Golden Rectangle.

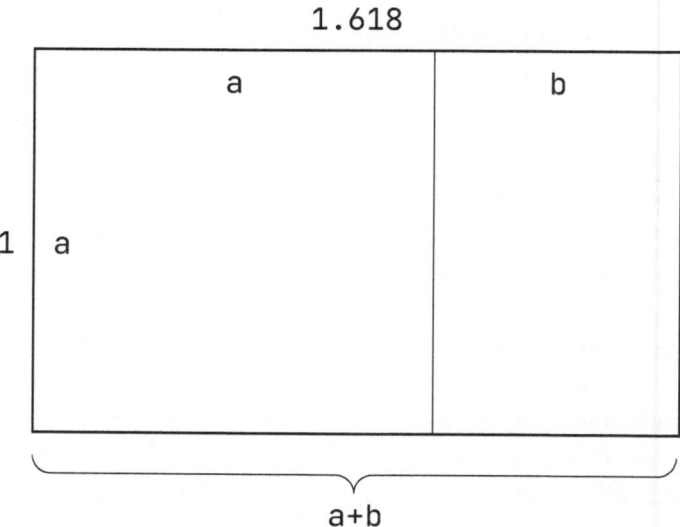

A shortcut to better remember the Golden Ratio is through the Fibonacci sequence. Instead of remembering 1 : 1.618033, we can simply memorise 3 : 5 or 5 : 8.

There are more reasons why the Golden ratio and, thereby, the Fibonacci numbers, play an important role in nature. Sunflowers, for example, grow their seeds in a spiral pattern (see Figure 64). To understand this relationship we have to look at how circular seeds are arranged in general.

We can focus our attention on the angle r at which the seeds are arranged. The left column in Figure 65 shows the first seed. The second seed is then placed at an angle r that is expressed as a fraction of a full circle. If the r is zero, then the second seed and all following seeds are placed on the same axis (see first row of Figure 65). If $r = \frac{1}{2}$, then the second seed is placed at half a circle from the first seed. The third seed is then placed after the first seed and thereby completes the circle. This results in all seeds placed along two spokes. If $r = \frac{1}{5}$, then the second seed is placed at one fifth of a circle from the first seed and so forth (see third row in Figure 65). The denominator seems to set the number of seed spokes. This method of arranging seeds is inefficient since it does not utilise the circular space of the flower. There is plenty of empty space.

Figure 64: The head of a sunflower. The number of spiralling spokes on the outer rim is always a Fibonacci number.

The magic starts to happen if we choose a number for r that cannot be expressed as a fraction. These are irrational numbers, as already discussed above. There are many irrational numbers, and π (pronounced /paɪ/)[22] might be one of the most well known examples. If we use $r = \frac{1}{\pi}$, then we get a far more efficient use of space (see Figure 66). There are other irrational numbers for r we could use, such as $\sqrt{2}$ or Euler's number.

The question arises as to what the best irrational number is for the arrangement of sunflower seeds, and the answer is $\phi = \frac{1+\sqrt{5}}{2}$, the Golden Ratio (see Figure 67). Despite the more organic placement of the seeds, we can still see spokes that spiral towards the centre. Moreover, the number of spokes at the rim of the sunflower is always a Fibonacci number. Figure 67 has 55 spokes.

But why would ϕ be a better irrational number compared to π? To answer this question we need an indicator for how irrational an irrational number is. Meaning, we need to quantify how well an irrational number can be approximated by a fraction. Let's start with π. We can approximate $\pi = 3.14159265359...$ by

$$\pi = 3 + \text{ a bit} \qquad (26)$$

"A bit" has to be less than 1 since otherwise π would be 4. Hence we can

[22]The 26 letters of the alphabet are insufficient for the 44 sounds known to the English language. The International Phonetic Alphabet is used to describe all the sounds of human oral languages.

Figure 65: Arrangement of sunflower seeds with the angle r as a fraction of a circle ($r = 0$, $r = \frac{1}{2}$ and $r = \frac{1}{5}$).

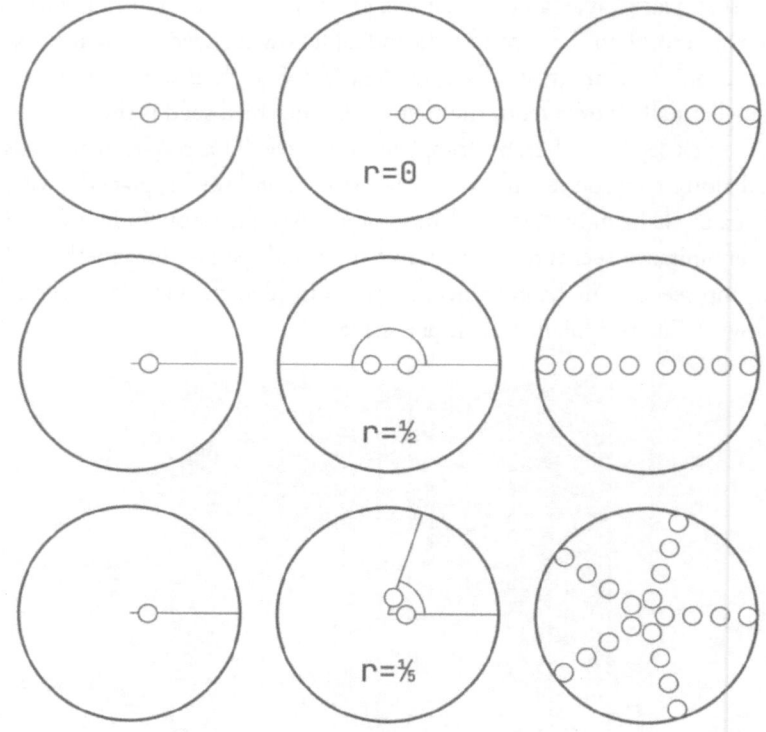

approximate 0.14159265359 as a fraction in the range of $\frac{1}{0<n<1}$:

$$\frac{1}{0.14159265359} = 7.062513305920733 \tag{27}$$

Therefore we can write:

$$\pi = 3 + \frac{1}{7 + \text{ a bit}} \tag{28}$$

We can then approximate 0.062513305920733 as a fraction:

$$\frac{1}{0.062513305920733} = 15.996594409324665 \tag{29}$$

Which we can then put back into the formula:

$$\pi = 3 + \cfrac{1}{7 + \cfrac{1}{15 + \text{ a bit}}} \tag{30}$$

We can then again approximate 0.996594409324665 as fraction:

$$\frac{1}{0.996594409324665} = 1.003417228356361 \tag{31}$$

$$\pi = 3 + \cfrac{1}{7 + \cfrac{1}{15 + \cfrac{1}{1 + \text{ a bit}}}} \tag{32}$$

Which leads us to:

$$\frac{1}{0.003417228356361} = 292.634818547771301 \tag{33}$$

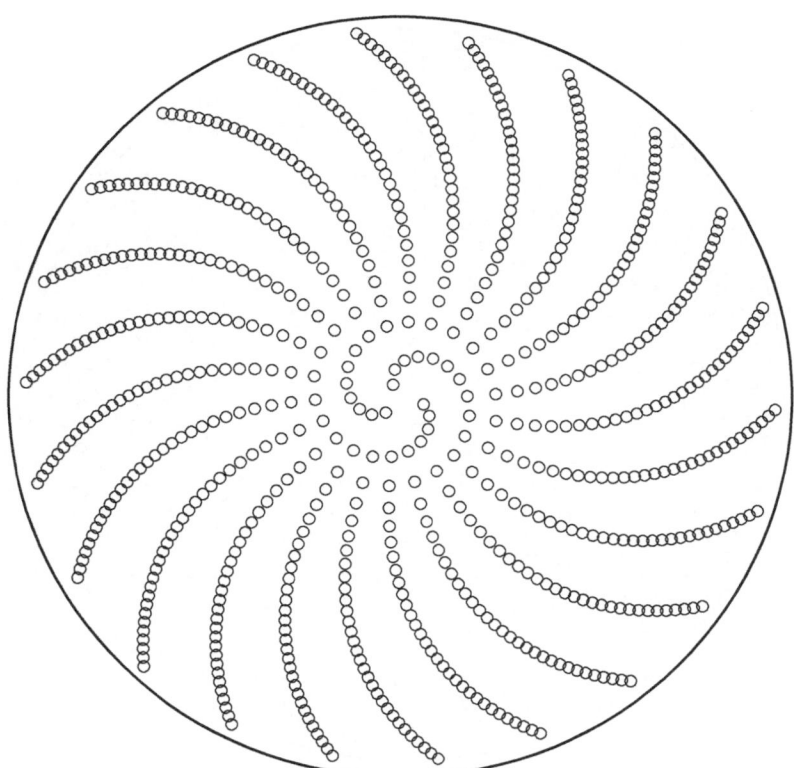

Figure 66: Arrangement of sunflower seeds with the angle $r = \frac{1}{\pi}$.

Which we can again enter into the formula:

$$\pi = 3 + \cfrac{1}{7 + \cfrac{1}{15 + \cfrac{1}{1 + \cfrac{1}{292 + \text{a bit}}}}} \tag{34}$$

Expressing an irrational number in this way is called a continued fraction, and it can be continued indefinitely. At this step already we notice that the change to approximate π is only $\frac{1}{292}$, which means that the approximation in the previous step was already very good. π can be approximated with this method reasonably well.

The most irrational irrational number, let's call it x, would then be a number that could be least approximated through a continued fraction. This means that at every step the "a bit" would not become small, such as $\frac{1}{292}$. The irrational number $\sqrt{2}$, for example, can be approximated as:

$$\sqrt{2} = 1 + \cfrac{1}{2 + \cfrac{1}{2 + \cfrac{1}{2 + \cfrac{1}{2 + ...}}}} \tag{35}$$

This is already an improvement since the increase in precision at every step remains constant. But it could even be more irrational. Ideally, the

Figure 67: Arrangement of sunflower seeds with the angle $r = \frac{1}{\phi}$.

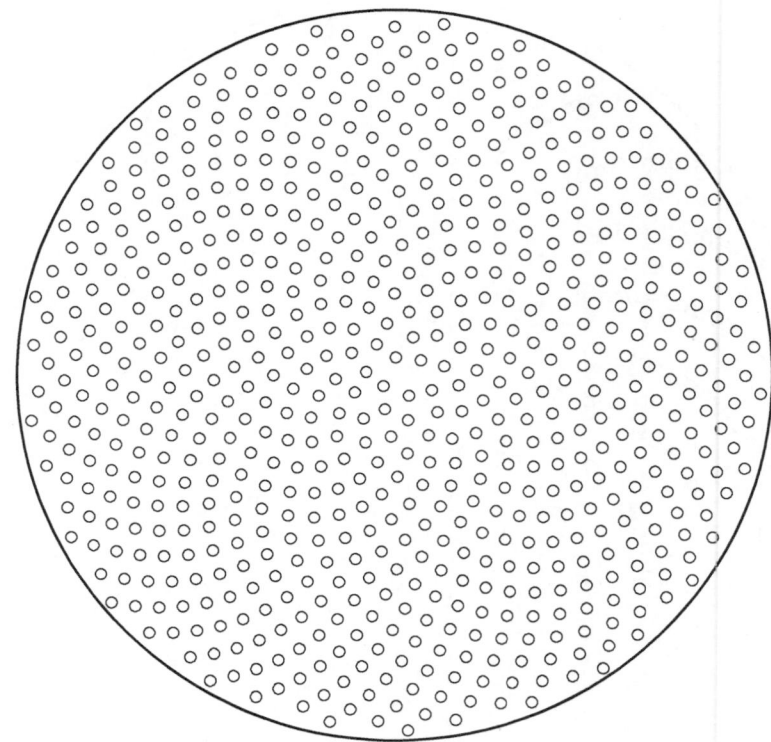

continued fraction would look like this:

$$x = 1 + \cfrac{1}{1 + \cfrac{1}{1 + \cfrac{1}{1 + \cfrac{1}{1 + \ldots}}}} \qquad (36)$$

To find x we need to take advantage of the fact that the highlighted part of Equation 37 below is self similar. The continued fraction is the same as itself.

$$x = 1 + \cfrac{1}{1 + \cfrac{1}{1 + \cfrac{1}{1 + \cfrac{1}{1 + \ldots}}}} \qquad (37)$$

We can replace the highlighted part of the formula with x itself, leading us to:

$$x = 1 + \frac{1}{x} \qquad (38)$$

To solve this equation we are going to multiply both sides by x:

$$x^2 = x + 1 \qquad (39)$$

And then rearrange it:

$$x^2 - x - 1 = 0$$
$$(x - \frac{1}{2})^2 - \frac{1}{4} - 1 = 0$$
$$(x - \frac{1}{2})^2 = \frac{5}{4}$$
$$x - \frac{1}{2} = \pm\frac{\sqrt{5}}{2} \tag{40}$$
$$x = \frac{1}{2} \pm \frac{\sqrt{5}}{2}$$
$$x = \frac{1 \pm \sqrt{5}}{2}$$

From Equation 24 we can see that this is exactly ϕ and hence $x = \phi$. Which means that ϕ is the most irrational of irrational numbers because it can be least approximated by a continued fraction. Calling somebody "as rational as the golden ratio" is thereby the most efficient and subtle insult.

Figure 68: The LEDs in modern traffic lights do unfortunately not follow a Fibonacci pattern.

Bonus

The Fibonacci sequence is not the only way to subdivide a rectangle, as seen in Figure 61. There are even better ways to do this. Let's first consider subdividing a square. If we fold a square piece of paper, we get a sheet with a proportion of 1:2. If we fold it again, we get back to a ratio of 1:2:

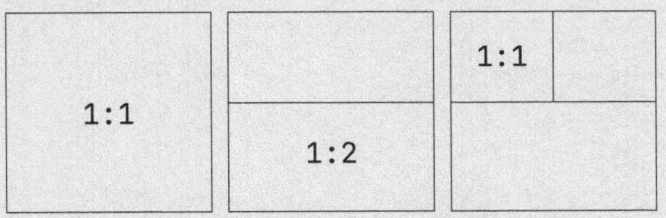

If we use the irrational number $\sqrt{2}$ as the ratio for our paper sheet then folding it will maintain the ratio of $1 : \sqrt{2}$:

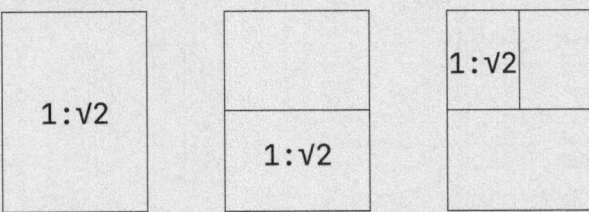

This is the basis for the DIN A standard used for paper. If we fold a piece of A4 paper, we get an A5 paper. While maintaining a ratio when folding is great, it does not yet provide us with an actual size. It is just a proportion. A0 anchors not only the ratio, but also the absolute size to 841 x 1189 mm, which is approximately one m^2 of paper.

Program

The Fibonacci sequence is another great example of how to use recursion in programming (see Listing 17).

```python
def fibonacci(n):
    if n <= 1:
        return n
    else:
        return (fibonacci(n-1) + fibonacci(n-2))

# set the number of terms to be calculated
n_terms = 10

print("Fibonacci sequence:")
for i in range(n_terms):
    print(fibonacci(i))
```

Listing 17: Algorithm to find the Fibonacci sequence.

Swimming the Fibonacci Sequence

With this in-depth understanding of the complexities associated with the Fibonacci sequence, we can generate a training program that is easy to remember and that will keep our mind busy.

A Python program that generates a swiML program at your desired lengths is even more useful (see Listing 18).

You can download this Python program from our repository.

```python
import swiML as swiML


def fibonacci(n):
    fib_sequence = [1, 1]   # Initialize with the first two terms

    # Generate Fibonacci sequence up to n terms
    for i in range(2, n):
        next_term = fib_sequence[-1] + fib_sequence[-2]
        fib_sequence.append(next_term)
    return fib_sequence

def create_swiML_instructions():
    my_instruction_list=[]
    j=0
    # main loop to create <instruction> elements based on the fibonacci
    ↪ array
    while j<number:
        if full_array[j]%2==0:
            current_stroke=even_stroke
        else:
            current_stroke=uneven_stroke
        my_instruction_list.append(swiML.Instruction(
            length=('lengthAsLaps',full_array[j]),
            stroke=('standardStroke',current_stroke),
            rest=('afterStop','PT0M15S')
        ))
```

```python
26          j+=1
27      return my_instruction_list
28
29  #writing the swiML program to disk
30  def write_program():
31      # warm up instructions
32      warmUp=swiML.Instruction(
33          length=('lengthAsDistance',400),
34          stroke=('standardStroke','any'),
35          intensity=('startIntensity',('zone','easy')),
36      )
37
38      # warm down instruction
39      warmDown=swiML.Instruction(
40          length=('lengthAsDistance',400),
41          stroke=('standardStroke','any'),
42          intensity=('startIntensity',('zone','easy')),
43      )
44      # the create the main instructions
45      myInstructions=[swiML.SegmentName('Warm
    ↪  Up'),warmUp,swiML.SegmentName('Fibonacci set')]
46      myInstructions.extend(create_swiML_instructions())
47      myInstructions.extend([swiML.SegmentName('Warm down'),warmDown])
48
49      # assemble the description of the swimming program
50      description_text="Swim the first "+str(number)+" terms of the
    ↪  Fibonacci sequence."
51
52      # create the program
53      simpleProgram=swiML.Program(
54          title='Fibonacci Sequence',
55          author=[('firstName','Christoph'),('lastName','Bartneck')],
56          programDescription=description_text,
57          poolLength='25',
58          creationDate='2024-08-22',
59          lengthUnit='meters',
60          hideIntro=False,
61          swiMLVersion='latest',
62          instructions=myInstructions
63      )
64      # write swiML XML to file
65      swiML.writeXML('patterns/fibonacci/fibonacci.xml',simpleProgram)
66
67  # counting up the number
68  number = 9
69  even_stroke="notFreestyle"
70  uneven_stroke="freestyle"
71
72  # create an array of the fibonacci sequence and write the program
73  full_array=fibonacci(number)
74  write_program()
```

Listing 18: Python program to create swiML program.

On line 70 we define the number of terms of the Fibonacci sequence.
The <length> of the program is defined in laps. This means that you can

influence the total length of the program by adjusting the pool length in line 59. On lines 71 and 72 you define the strokes for even and uneven terms.

Figure 69: The Fibonacci sequence program. The first term is 0, but it has been excluded from this program. You can download this swimming program as a PDF from our repository.

Warm Up	
400 Any Easy	1
Fibonacci set	
1 laps FR ↻0:15	2
1 laps FR ↻0:15	3
2 laps Not FR ↻0:15	4
3 laps FR ↻0:15	5
5 laps FR ↻0:15	6
8 laps Not FR ↻0:15	7
13 laps FR ↻0:15	8
21 laps FR ↻0:15	9
34 laps Not FR ↻0:15	10
Warm down	
400 Any Easy	11

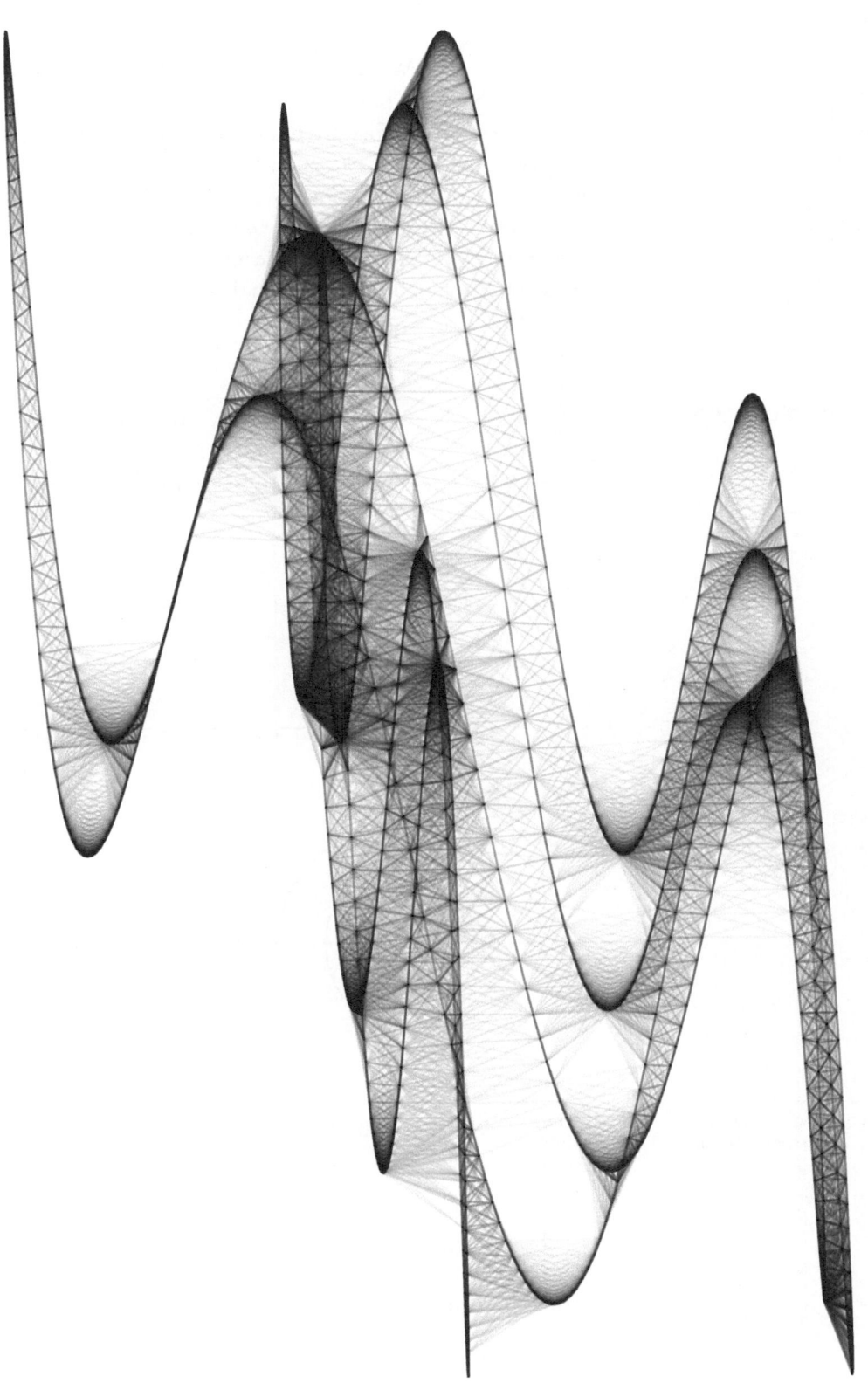

Figure 70: Colombian mathematician Bernardo Recaman Santos (∗1954) (source: Bernardo Recaman Santos).

Figure 71: The first five steps in the Recaman sequence.

Recaman's Sequence

Bernardo Recaman Santos (see Figure 70) created this extremely simple and beautiful sequence. It is best explained by using the natural number line (see Figure 71). We start at the number 0 and move one step. We cannot move backwards, since this would be a negative number. We therefore have to move forward to number 1 (see Figure 71i). Next, we move by two steps. Again, we cannot go backward, so we move forward to 3 (see Figure 71ii). Now we move by three steps. We cannot go backward since we already visited the number 0. Hence we have to move forward to 6 (see Figure 71iii). Next, we move four steps. This time we can move backward to 2 (see Figure 71iv). From there we move forward again to 7 (see Figure 71v)

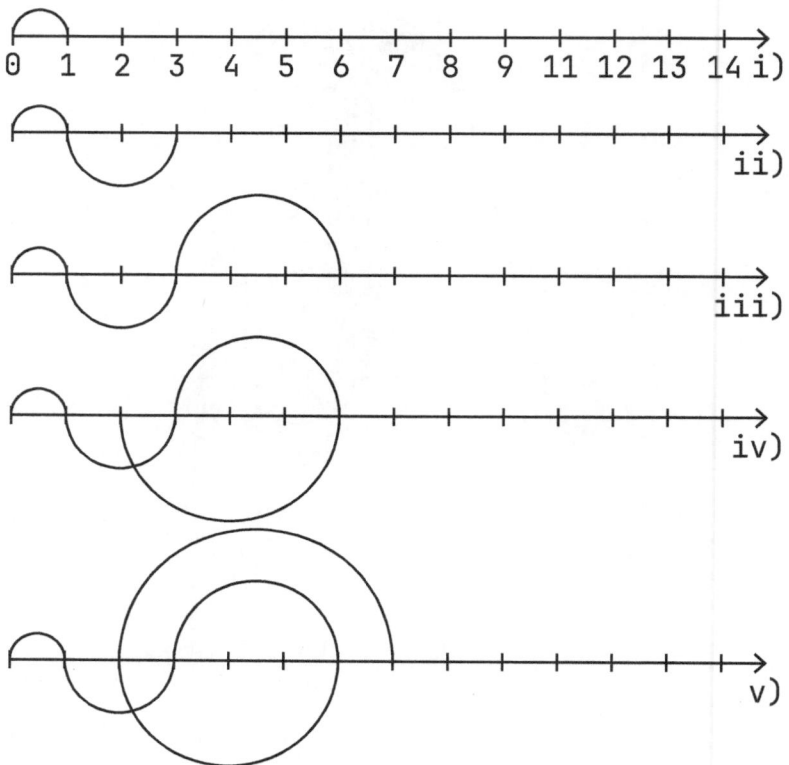

This sequence is registered with the On-Line Encyclopedia of Integer Sequences as A005132 (https://oeis.org/A0051 32)

Formally, the Recaman sequence r is defined as:

$$r_n = \begin{cases} 0 & \text{if n=0} \\ r_{n-1} & \text{if } r_{n-1} - n > 0 \text{ and is not already in the sequence} \\ r_{n+1} & \text{otherwise} \end{cases} \quad (41)$$

The resulting sequence is:

$$0, 1, 3, 6, 2, 7, 13, 20, 12, 21, 11, 22, 10, 23, 9, 24, 8, 25, \ldots$$

The open question is if all natural numbers will occur in this sequence. While no formal proof is available yet, a large number of terms have been tried. 10^{230} of terms, to be precise. The evidence so far seems to suggest that all natural numbers will occur in this sequence, but trying is not the same as proving.

What we can be certain of is that a visual representation of this sequence looks amazing (see Figure 72). It is also shown on the front cover of this book.

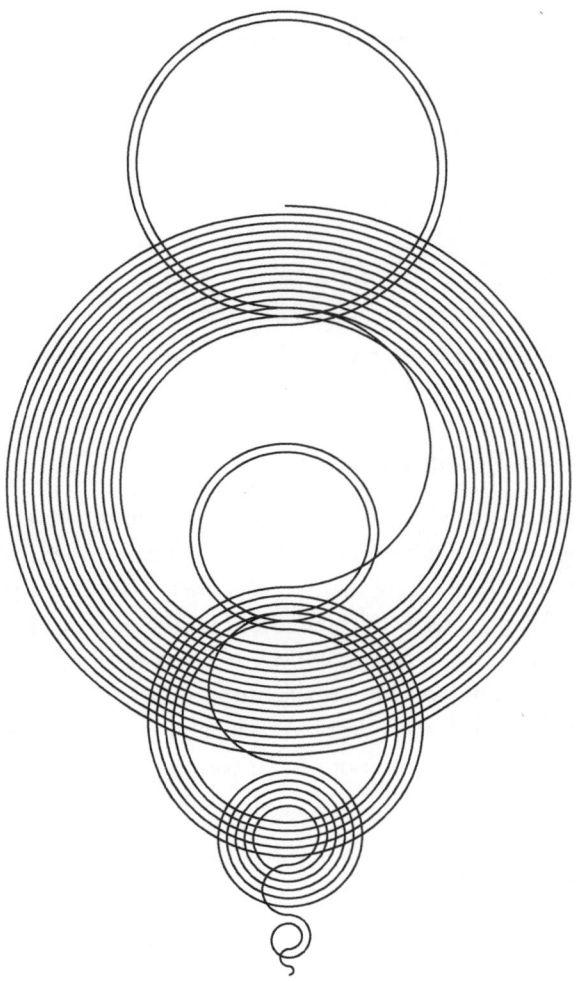

Figure 72: Visualisation of the Recaman sequence.

Program

The Recaman sequence can be calculated using the following Python program:

```python
# Initialize the sequence with the first term
seq = [0]
steps = []

# Calculate the remaining terms
for i in range(1, 100):
    # Calculate the next term using the recurrence relation
    next_term = seq[-1] - i

```

```
10      # If the next term is negative or has already appeared in the
        ↪ sequence,
11      # use the formula with addition instead
12      if next_term in seq or next_term < 0:
13          next_term = seq[-1] + i
14
15      # Determine whether the step is forward or backward
16      step = 'forward' if next_term > seq[-1] else 'backward'
17      steps.append(step)
18
19      # Add the next term to the sequence
20      seq.append(next_term)
21
22  # Output the sequence and steps
23  print('Sequence:', seq)
24  print('Steps:', steps)
```

Listing 19: Algorithm to find the Recaman's sequence.

Swimming the Recaman sequence

It would be convenient to be able to swim any number of terms of the Recaman sequence. For this we need a little Python program (see Listing 20). On line 71 we define how many terms we are looking for. On line 3 we first find the terms with the same function used in Listing 19. The function in line 20 then creates swiML instructions based on the sequence. The numerical value of the term defines the lap count (line 24). If this term moves forward, we swim freestyle, and if it moves backward, we swim backstroke (line 25). Notice that we start the program not at zero, but at one lap.

You can download this Python program from our repository.

```
1   import swiML as swiML
2   def find_recaman(terms):
3       for i in range(1, terms):
4           # Calculate the next term using the recurrence relation
5           next_term = seq[-1] - i
6           # If the next term is negative or has already appeared in the
            ↪ sequence,
7           # use the formula with addition instead
8           if next_term in seq or next_term < 0:
9               next_term = seq[-1] + i
10          # Determine whether the step is forward or backward
11          step = 'freestyle' if next_term > seq[-1] else 'backstroke'
12          steps.append(step)
13          # Add the next term to the sequence
14          seq.append(next_term)
15
16  def create_swiML_instructions():
17      my_instruction_list=[]
18      for i in range(1, nr_terms+1):
19          my_instruction_list.append(swiML.Instruction(
20              length=('lengthAsLaps',seq[i]),
21              stroke=('standardStroke',steps[i]),
```

```
22              rest=('afterStop','PT0M15S')
23          ))
24          i+=1
25      return my_instruction_list
26
27  #writing the swiML program to disk
28  def write_program():
29      # warm up instructions
30      warmUp=swiML.Instruction(
31          length=('lengthAsDistance',400),
32          stroke=('standardStroke','any'),
33          intensity=('startIntensity',('zone','easy')),
34      )
35      # warm down instruction
36      warmDown=swiML.Instruction(
37          length=('lengthAsDistance',200),
38          stroke=('standardStroke','any'),
39          intensity=('startIntensity',('zone','easy')),
40      )
41      # assembly of the main instructions
42      myInstructions=[swiML.SegmentName('Warm
        ↪ Up'),warmUp,swiML.SegmentName('Recaman set')]
43      myInstructions.extend(create_swiML_instructions())
44      myInstructions.extend([swiML.SegmentName('Warm down'),warmDown])
45      # assemble the description of the swimming program
46      description_text="Swim the first "+str(nr_terms)+" terms of the
        ↪ Recaman sequence. "
47      # create the program
48      simpleProgram=swiML.Program(
49          title='Recaman Sequence',
50          author=[('firstName','Christoph'),('lastName','Bartneck')],
51          programDescription=description_text,
52          poolLength='25',
53          creationDate='2024-08-22',
54          lengthUnit='meters',
55          hideIntro=False,
56          swiMLVersion='latest',
57          instructions=myInstructions
58      )
59      # write swiML XML to file
60      swiML.writeXML('patterns/recaman/recaman-program.xml',simpleProgram)
61  # define the number of terms
62  nr_terms=10
63  # Initialize the sequence with the first term
64  seq = [0]
65  steps = []
66  # find the terms
67  find_recaman(nr_terms+2)
68  # write the swiML program
69  write_program()
```

Listing 20: Python program to create the Recaman program.

Here is the resulting Recaman program for ten terms, excluding the first first term of 0. In a 25 meter pool this adds up to 3,000 meters.

Figure 73: Swim the Recaman sequence.
You can download this swimming program
as a PDF from our repository.

Warm Up	
400 Any Easy	1
Recaman set	
1 laps FR ↻0:15	2
3 laps FR ↻0:15	3
6 laps BK ↻0:15	4
2 laps FR ↻0:15	5
7 laps FR ↻0:15	6
13 laps FR ↻0:15	7
20 laps BK ↻0:15	8
12 laps FR ↻0:15	9
21 laps BK ↻0:15	10
11 laps FR ↻0:15	11
Warm down	
200 Any Easy	12

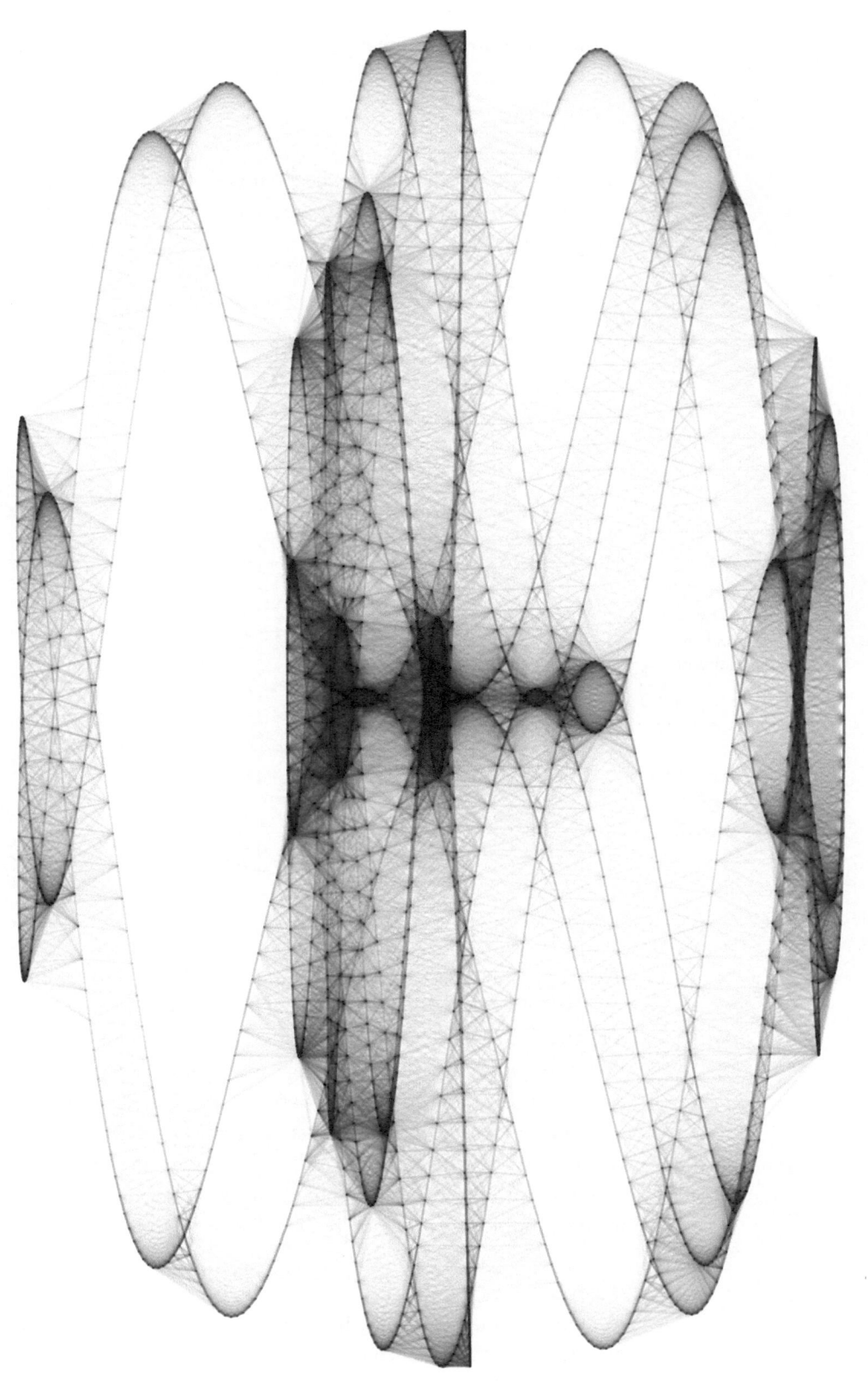

Pi

The number Pi, commonly represented with the Greek letter π, is one of the most important numbers in mathematics. It is defined as the ratio of a circle's circumference C to its diameter d:

$$\pi = \frac{C}{d} \tag{42}$$

Based on this we can calculate the circumference of any circle. For a circle with diameter of 2, the radius is 1 since $r = \frac{d}{2}$. For a circle of radius 1 the circumference is:

$$\begin{aligned}
C &= d \times \pi \\
&= 2 \times r \times \pi \\
&= 2 \times 1 \times \pi \\
&= 2\pi
\end{aligned} \tag{43}$$

The area of a circle with $r = 1$ is:

$$\begin{aligned}
A &= r^2 \times \pi \\
&= 1^2 \times \pi \\
&= \pi
\end{aligned} \tag{44}$$

Figure 74: Radius, diameter, area and circumference of a circle with the radius of one.

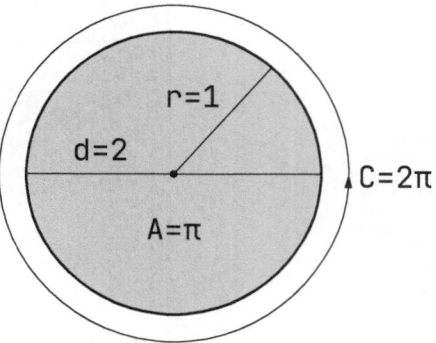

This relationship is a constant, meaning that it is the same for circles of any size. Still, it is an irrational number. This means it cannot be expressed as a fraction of two integer numbers. Its digits will go on forever, and they will never enter a permanently repeating pattern. An example of such a repeating pattern would be $\frac{1}{3} = 0.3333333...$ or $\frac{9}{11} = 0.8181818181...$. The digits for π do not have such patterns. The digits are seemingly random. What we do assume is that all numbers occur equally often. A zero seems to occur just as often as a five. We do not, however, have a formal proof for this conjecture. The first 20 digits are:

$$3.14159265358979323846...$$

Amongst the mathematically interested, it is a common challenge to memorise digits of π. The current world record is around 70,000 digits. Computers have pushed this to 100 trillion digits in 2022. Intensive calculations and large memory banks are necessary for the calculation of π. Its digits cannot be predicted, and they must be calculated.

How many digits should a swimmer remember? If we assume a pool length of 50 meters, then we can calculate the circumference of a circle with $r = 50$ (see Figure 75) following the calculation in Equation 45.

$$C = 2\pi r$$
$$C = 2 \times \pi \times 50 = 314.16$$

(45)

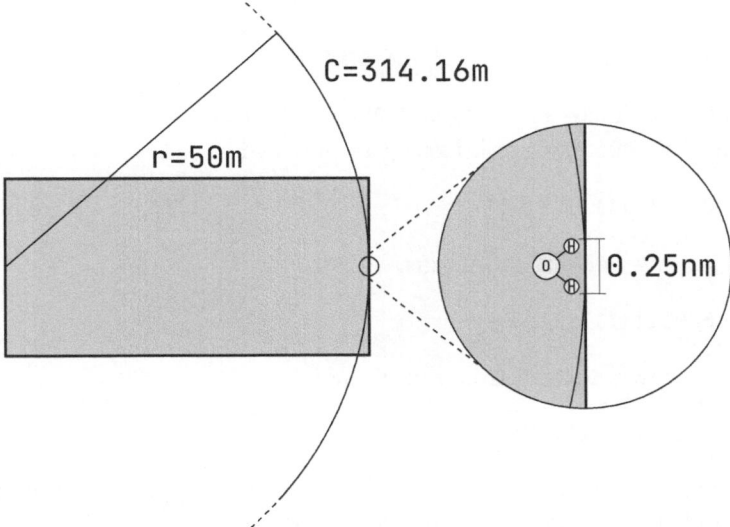

Figure 75: A water molecule in a 50 × 25 meter swimming pool.

A water molecule has a diameter of 0.27 nanometers. Increasing the number of digits from 12 to 13 for π changes the calculation of the circumference by less than the size of a water molecule. Hence a precision of 13 digits should be the smallest meaningful approximation of π for a swimmer. Hence the swimmers' π is:

3.1415926535897

You can download the Microsoft Excel file here to try out this error:

Bonus

Microsoft Excel works with 15 digit precision. For Excel π is exactly 3.14159265358979. This will be enough precision for almost all swimming related calculations, but it is surprising to see how easily errors occur in this widely spread software. Even the simple mathematical equation:

$$1 + x - 1 = x$$

cannot be guaranteed in Excel. We can try this out ourselves. First, enter 1/9000 into a cell. Excel presents us with:

`0.000111111111111111`

If we now add 1 to this we get:

`1.000111111111110000`

If we now subtract 1 from this we get:

`0.000111111111111173`

Adding 1 and subtracting 1 should bring us back to the original number. But it does not. Here is what the calculation looks like in Excel.

	A	B
1	Formula	Result
2	B2=1/9000:	0.000111111111111111
3	B3=B2+1:	1.000111111111110000
4	B4=B3-1:	0.000111111111111173
5		

Swimming Pi

The digits of π are random and cannot be predicted. We have to calculate each one of them. To do so, we can import the Python mpmath library[23] to speed up the process (see Listing 21). We can then set the number of decimal places in line 4, convert the digits first into a string in line 5 before converting the digits into an array.

[23] https://mpmath.org

```python
import mpmath

def pi_digits(n):
    mpmath.mp.dps = n + 1  # Set the number of decimal places
    pi_str = str(mpmath.pi)  # Get Pi as a string with desired precision
    pi_digits_array = [int(digit) for digit in pi_str.replace('.', '')]
    ↪  # Convert Pi string to array of digits
    return pi_digits_array
```

```
8
9    # Example usage
10   n = 20   # Number of digits desired
11   pi_array = pi_digits(n)
12   print(pi_array)
```

Listing 21: Listing the digits of π.

Using this function, we can now construct a swimming program of any length using the digits of π. If we use the 13 digits of the π for swimmers, we get a program that is 4,000 meters in a 50 meters pool. The digits of π determine the length of each instruction (line 22). Uneven digits are swum in Freestyle while even numbers in notFreestyle (line 23). Short distances are swum at high intensity, while long distances are swum at low intensity (line 24). Figure 76 shows the whole training program. We can easily modify this program to adapt it to a 25 meter pool (line 59), or change the number of digits (line 70).

You can download this Python program from our repository.

```
1    import swiML as swiML, mpmath, math
2
3    def pi_digits(n):
4        # Set the number of decimal places
5        mpmath.mp.dps = n + 1
6        # Get Pi as a string with desired precision
7        pi_str = str(mpmath.pi)
8        # Convert Pi string to array of digits
9        pi_digits_array = [int(digit) for digit in pi_str.replace('.', '')]
10       return pi_digits_array
11
12   def create_swiML_instructions():
13       my_instruction_list=[]
14       for i in range(0, digits):
15           if pi_array[i]%2==0:
16               current_stroke="notFreestyle"
17           else:
18               current_stroke="freestyle"
19           intensity_list=["max","racePace","endurance","threshold","easy"]
20           current_intensity=intensity_list[math.floor(pi_array[i]/2)]
21           my_instruction_list.append(swiML.Instruction(
22                   length=('lengthAsLaps',pi_array[i]),
23                   stroke=('standardStroke',current_stroke),
24                   intensity=('startIntensity',('zone',current_intensity)),
25                   rest=('afterStop','PT0M15S')
26           ))
27           i+=1
28       return my_instruction_list
29
30   #writing the swiML program to disk
31   def write_program():
32       # warm up instructions
33       warmUp=swiML.Instruction(
34           length=('lengthAsDistance',400),
35           stroke=('standardStroke','any'),
36           intensity=('startIntensity',('zone','easy')),
37       )
```

```
38      # warm down instruction
39      warmDown=swiML.Instruction(
40          length=('lengthAsDistance',200),
41          stroke=('standardStroke','any'),
42          intensity=('startIntensity',('zone','easy')),
43      )
44      # assembly of the main instructions
45      myInstructions=[swiML.SegmentName('Warm
        ↪  Up'),warmUp,swiML.SegmentName('Pi set')]
46      myInstructions.extend(create_swiML_instructions())
47      myInstructions.extend([swiML.SegmentName('Warm down'),warmDown])
48
49      # assemble the description of the swimming program
50      description_text="Swim the first "+str(digits-1)+" digits of Pi
        ↪  while increasing the intensity for shorter distances. "
51
52      # create the program
53      simpleProgram=swiML.Program(
54          title='Palatial Pi Program',
55          author=[('firstName','Christoph'),('lastName','Bartneck')],
56          programDescription=description_text,
57          poolLength='50',
58          creationDate='2024-08-22',
59          lengthUnit='meters',
60          hideIntro=False,
61          swiMLVersion='latest',
62          instructions=myInstructions
63      )
64      # write swiML XML to file
65      swiML.writeXML('patterns/pi/pi-program.xml',simpleProgram)
66
67  # Number of digits desired
68  digits = 14
69  pi_array = pi_digits(digits)
70  # write the swiML program
71  write_program()
```

Listing 22: Creating π swiML programs.

Warm Up	
400 Any Easy	1
Pi set	
3 laps FR ⟳0:15 Race Pace	2
1 laps FR ⟳0:15 Max	3
4 laps Not FR ⟳0:15 Endurance	4
1 laps FR ⟳0:15 Max	5
5 laps FR ⟳0:15 Endurance	6
9 laps FR ⟳0:15 Easy	7
2 laps Not FR ⟳0:15 Race Pace	8
6 laps Not FR ⟳0:15 Threshold	9
5 laps FR ⟳0:15 Endurance	10
3 laps FR ⟳0:15 Race Pace	11
5 laps FR ⟳0:15 Endurance	12
8 laps Not FR ⟳0:15 Easy	13
9 laps FR ⟳0:15 Easy	14
7 laps FR ⟳0:15 Threshold	15
Warm down	
200 Any Easy	16

Figure 76: Swim the digits of π sequence. You can download this swim program from our repository.

In case you want to swim more digits of π but you do not have a computer at hand, here are a couple of pages of π for you:

3.

```
1415926535 8979323846 2643383279 5028841971 6939937510
5820974944 5923078164 0628620899 8628034825 3421170679
8214808651 3282306647 0938446095 5058223172 5359408128
4811174502 8410270193 8521105559 6446229489 5493038196
4428810975 6659334461 2847564823 3786783165 2712019091
4564856692 3460348610 4543266482 1339360726 0249141273
7245870066 0631558817 4881520920 9628292540 9171536436
7892590360 0113305305 4882046652 1384146951 9415116094
3305727036 5759591953 0921861173 8193261179 3105118548
0744623799 6274956735 1885752724 8912279381 8301194912
9833673362 4406566430 8602139494 6395224737 1907021798
6094370277 0539217176 2931767523 8467481846 7669405132
0005681271 4526356082 7785771342 7577896091 7363717872
1468440901 2249534301 4654958537 1050792279 6892589235
4201995611 2129021960 8640344181 5981362977 4771309960
5187072113 4999999837 2978049951 0597317328 1609631859
5024459455 3469083026 4252230825 3344685035 2619311881
7101000313 7838752886 5875332083 8142061717 7669147303
5982534904 2875546873 1159562863 8823537875 9375195778
1857780532 1712268066 1300192787 6611195909 2164201989

3809525720 1065485863 2788659361 5338182796 8230301952
0353018529 6899577362 2599413891 2497217752 8347913151
5574857242 4541506959 5082953311 6861727855 8890750983
8175463746 4939319255 0604009277 0167113900 9848824012
8583616035 6370766010 4710181942 9555961989 4676783744
9448255379 7747268471 0404753464 6208046684 2590694912
9331367702 8989152104 7521620569 6602405803 8150193511
2533824300 3558764024 7496473263 9141992726 0426992279
6782354781 6360093417 2164121992 4586315030 2861829745
5570674983 8505494588 5869269956 9092721079 7509302955
3211653449 8720275596 0236480665 4991198818 3479775356
6369807426 5425278625 5181841757 4672890977 7727938000
8164706001 6145249192 1732172147 7235014144 1973568548
1613611573 5255213347 5741849468 4385233239 0739414333
4547762416 8625189835 6948556209 9219222184 2725502542
5688767179 0494601653 4668049886 2723279178 6085784383
8279679766 8145410095 3883786360 9506800642 2512520511
7392984896 0841284886 2694560424 1965285022 2106611863
0674427862 2039194945 0471237137 8696095636 4371917287
4677646575 7396241389 0865832645 9958133904 7802759009

9465764078 9512694683 9835259570 9825822620 5224894077
2671947826 8482601476 9909026401 3639443745 5305068203
4962524517 4939965143 1429809190 6592509372 2169646151
5709858387 4105978859 5977297549 8930161753 9284681382
6868386894 2774155991 8559252459 5395943104 9972524680
8459872736 4469584865 3836736222 6260991246 0805124388
```

```
4390451244 1365497627 8079771569 1435997700 1296160894
4169486855 5848406353 4220722258 2848864815 8456028506
0168427394 5226746767 8895252138 5225499546 6672782398
6456596116 3548862305 7745649803 5593634568 1743241125
1507606947 9451096596 0940252288 7971089314 5669136867
2287489405 6010150330 8617928680 9208747609 1782493858
9009714909 6759852613 6554978189 3129784821 6829989487
2265880485 7564014270 4775551323 7964145152 3746234364
5428584447 9526586782 1051141354 7357395231 1342716610
2135969536 2314429524 8493718711 0145765403 5902799344
0374200731 0578539062 1983874478 0847848968 3321445713
8687519435 0643021845 3191048481 0053706146 8067491927
8191197939 9520614196 6342875444 0643745123 7181921799
9839101591 9561814675 1426912397 4894090718 6494231961

5679452080 9514655022 5231603881 9301420937 6213785595
6638937787 0830390697 9207734672 2182562599 6615014215
0306803844 7734549202 6054146659 2520149744 2850732518
6660021324 3408819071 0486331734 6496514539 0579626856
1005508106 6587969981 6357473638 4052571459 1028970641
4011097120 6280439039 7595156771 5770042033 7869936007
2305587631 7635942187 3125147120 5329281918 2618612586
7321579198 4148488291 6447060957 5270695722 0917567116
7229109816 9091528017 3506712748 5832228718 3520935396
5725121083 5791513698 8209144421 0067510334 6711031412
6711136990 8658516398 3150197016 5151168517 1437657618
3515565088 4909989859 9823873455 2833163550 7647918535
8932261854 8963213293 3089857064 2046752590 7091548141
6549859461 6371802709 8199430992 4488957571 2828905923
2332609729 9712084433 5732654893 8239119325 9746366730
5836041428 1388303203 8249037589 8524374417 0291327656
1809377344 4030707469 2112019130 2033038019 7621101100
4492932151 6084244485 9637669838 9522868478 3123552658
2131449576 8572624334 4189303968 6426243410 7732269780
2807318915 4411010446 8232527162 0105265227 2111660396

6655730925 4711055785 3763466820 6531098965 2691862056
4769312570 5863566201 8558100729 3606598764 8611791045
3348850346 1136576867 5324944166 8039626579 7877185560
8455296541 2665408530 6143444318 5867697514 5661406800
7002378776 5913440171 2749470420 5622305389 9456131407
1127000407 8547332699 3908145466 4645880797 2708266830
6343285878 5698305235 8089330657 5740679545 7163775254
2021149557 6158140025 0126228594 1302164715 5097925923
0990796547 3761255176 5675135751 7829666454 7791745011
2996148903 0463994713 2962107340 4375189573 5961458901
9389713111 7904297828 5647503203 1986915140 2870808599
0480109412 1472213179 4764777262 2414254854 5403321571
8530614228 8137585043 0633217518 2979866223 7172159160
```

```
7716692547  4873898665  4949450114  6540628433  6639379003
9769265672  1463853067  3609657120  9180763832  7166416274
8888007869  2560290228  4721040317  2118608204  1900042296
6171196377  9213375751  1495950156  6049631862  9472654736
4252308177  0367515906  7350235072  8354056704  0386743513
6222247715  8915049530  9844489333  0963408780  7693259939
7805419341  4473774418  4263129860  8099888687  4132604721
                                        .

5695162396  5864573021  6315981931  9516735381  2974167729
4786724229  2465436680  0980676928  2382806899  6400482435
4037014163  1496589794  0924323789  6907069779  4223625082
2168895738  3798623001  5937764716  5122893578  6015881617
5578297352  3344604281  5126272037  3431465319  7777416031
9906655418  7639792933  4419521541  3418994854  4473456738
3162499341  9131814809  2777710386  3877343177  2075456545
3220777092  1201905166  0962804909  2636019759  8828161332
3166636528  6193266863  3606273567  6303544776  2803504507
7723554710  5859548702  7908143562  4014517180  6246436267
9456127531  8134078330  3362542327  8394497538  2437205835
3114771199  2606381334  6776879695  9703098339  1307710987
0408591337  4641442822  7726346594  7047458784  7787201927
7152807317  6790770715  7213444730  6057007334  9243693113
8350493163  1284042512  1925651798  0694113528  0131470130
4781643788  5185290928  5452011658  3934196562  1349143415
9562586586  5570552690  4965209858  0338507224  2648293972
8584783163  0577775606  8887644624  8246857926  0395352773
4803048029  0058760758  2510474709  1643961362  6760449256
2742042083  2085661190  6254543372  1315359584  5068772460

2901618766  7952406163  4252257719  5429162991  9306455377
9914037340  4328752628  8896399587  9475729174  6426357455
2540790914  5135711136  9410911939  3251910760  2082520261
8798531887  7058429725  9167781314  9699009019  2116971737
2784768472  6860849003  3770242429  1651300500  5168323364
3503895170  2989392233  4517220138  1280696501  1784408745
1960121228  5993716231  3017114448  4640903890  6449544400
6198690754  8516026327  5052983491  8740786680  8818338510
2283345085  0486082503  9302133219  7155184306  3545500766
8282949304  1377655279  3975175461  3953984683  3936383047
4611996653  8581538420  5685338621  8672523340  2830871123
2827892125  0771262946  3229563989  8989358211  6745627010
2183564622  0134967151  8819097303  8119800497  3407239610
3685406643  1939509790  1906996395  5245300545  0580685501
9567302292  1913933918  5680344903  9820595510  0226353536
1920419947  4553859381  0234395544  9597783779  0237421617
2711172364  3435439478  2218185286  2408514006  6604433258
8856986705  4315470696  5747458550  3323233421  0730154594
0516553790  6866273337  9958511562  5784322988  2737231989
8757141595  7811196358  3300594087  3068121602  8764962867
```

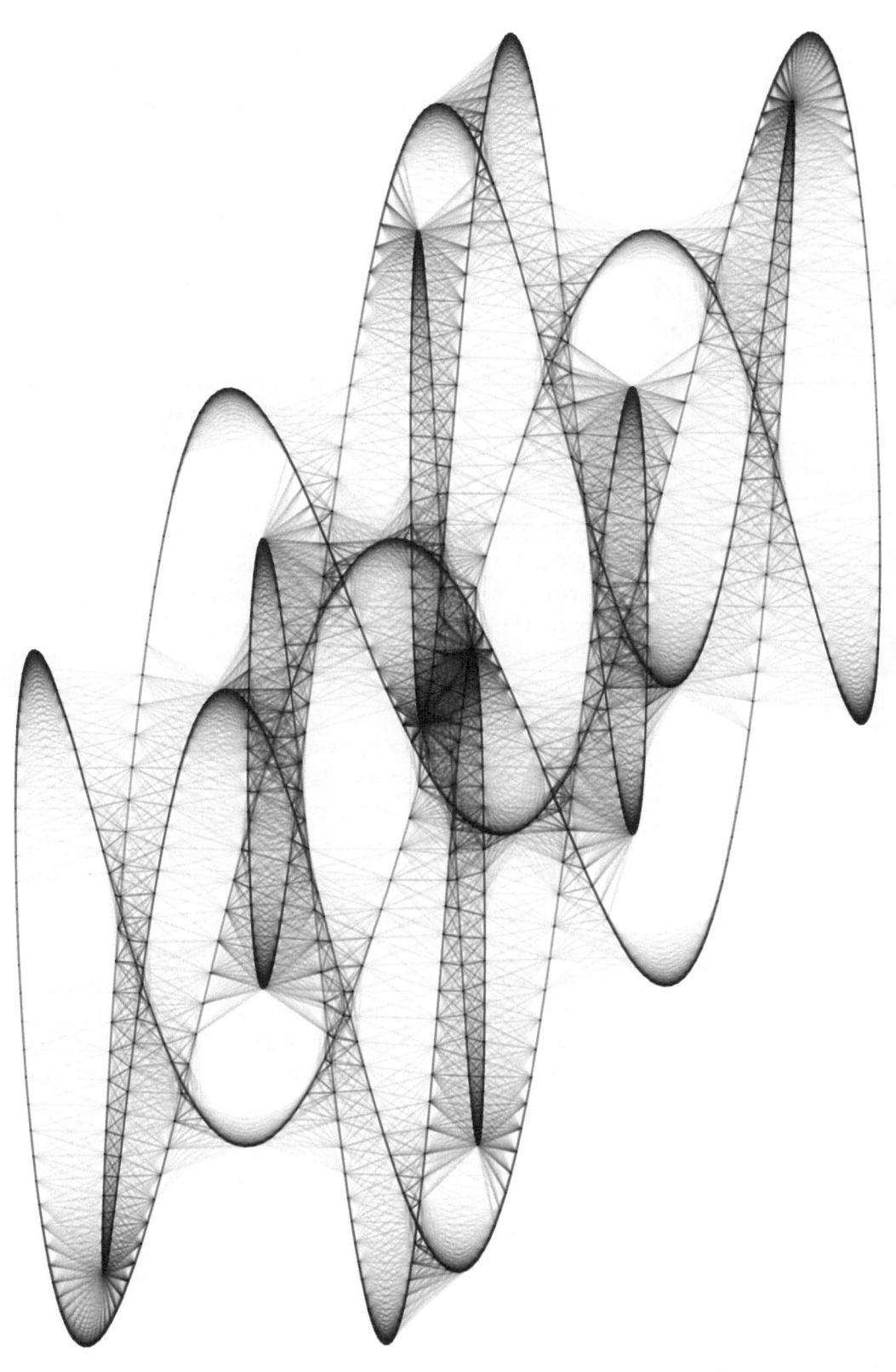

Roman Numerals

We have already encountered different numerical systems, such as the binary system (see page 63) or the hexadecimal system (see page 64). The latter used the letters A–F for the numerical values of 11–16. The Roman numeral system also uses letters for numerical values:

Table 6: Roman numerals.

I	1
V	5
X	10
L	50
C	100
D	500
M	1000

Figure 77: Roman emperor Gaius Julius Caesar (12 July 100 BC – 15 March 44 BC) (source: Ángel M. Felicísimo).

There is no letter for the numerical value of 2, 3 or 4. Nor for 8, 39 or 42. To express the numerical value of 2, the letter for 1 gets repeated twice: II. This works up to a repetition of three letters. 3 is expressed as III, 30 is XXX and 300 is CCC.

The expression for 4 introduces the subtractive connotation IV. By writing I before the letter V we express $5 - 1 = 4$. Nine is then expressed as IX and 39 is XXXIX. This is shorter than writing VIIII or XXVIIII.

Notice that Roman numerals are written from the largest to the smallest letter. 1111 is expressed as MCXI. When a smaller letter is located before a larger letter, it indicates the subtractive method. 999 is therefore CMXCIX. The largest number that can be expressed with this system is 3,999 as MMMCMXCIX.

In swiML we can easily switch the numeral system to Roman with the `<numeralSystem>` element. By default, it is set to decimal, but it can be set to Roman. In the future, we hope to support other numeral systems.

Swim Roman Numerals

Any swimming program can be expressed in Roman numerals. You only need to set the `<numeralSystem>` element to `roman`. The regular programs in chapter Regular Programs are a good starting point for such a conversion. As an example here, we focus on the birthdate of the Roman emperor Gaius Julius Caesar (see Figure 77). He was born on 12 July 100 BC. We can swim his birthday as:

	Warm up	
CD Any Easy		1
	First set	
XII × **C** FR @_1:45 60…90%		2
VII × **CC** FR @_1:45 Pads Pullbuoy		3
	Warm down	
C Any Easy		4

Figure 78: Swim the birthday of Gaius Julius Cesar. You can download this swim program from our repository.

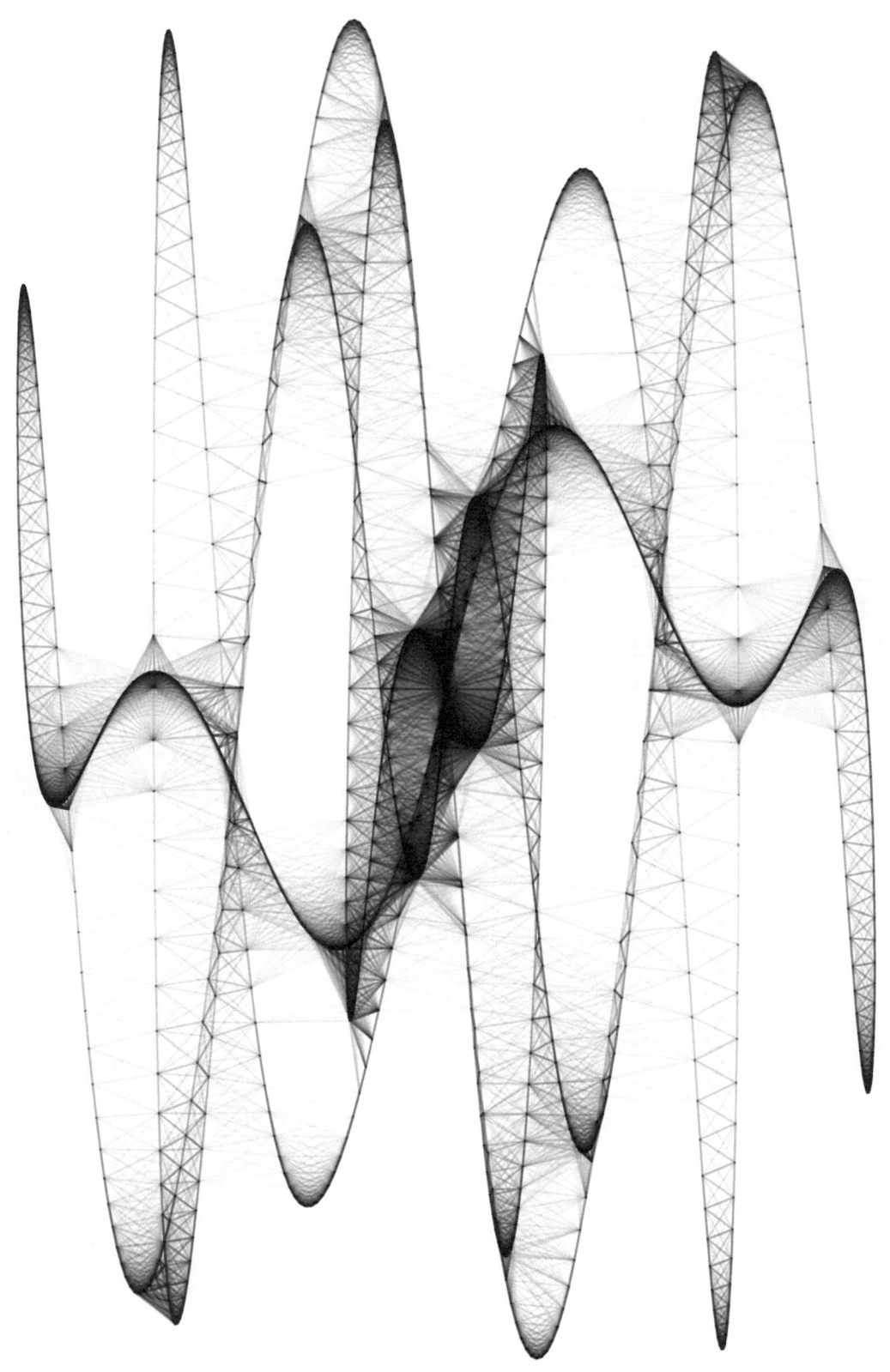

Prime Numbers

Prime numbers are natural numbers that are greater than 1 and that are only divisible by 1 and themselves. The number 11, for example, is greater than 1 and can only be divided by 1 and 11. Prime numbers play an important role in maths and computer science. There are infinitely many prime numbers, and finding them becomes increasingly difficult the larger the number becomes. Here are the first 20 prime numbers:

$$2, 3, 5, 7, 11, 13, 17, 19, 23, 29, 31, 37, 41, 43, 47, 53, 59, 61, 67, 71 \quad (46)$$

The trouble with primes is that they become increasingly rare and there is no easy way to predict their occurrence. In comparison, even numbers are frequent, and we can predict that every second number is even. Finding prime numbers is difficult, but we can approximate certain aspects of them. For example, the probability p that the number n is a prime number is approximately:

The sequence of prime numbers is registered with the On-Line Encyclopedia of Integer Sequences as A000040 (https://oeis.org/A000040).

$$p_n \approx \frac{1}{\log n} \quad (47)$$

To understand what a logarithm log is, we should start with its inversion, the exponentiation.

$$10^3 = 10 \times 10 \times 10 = 1000 \quad (48)$$

The logarithm of 1000 is then the number by which 10 has to be multiplied by itself to result in 1000. In our example in Equation 48 this would be 3. We can write this as:

$$\log 1000 = 3 \quad (49)$$

So far we only considered the exponentiation and logarithm with a base of 10. There are other popular bases, such as 2 or Euler's constant e, named after the German mathematician Leonhard Euler (see Figure 79). The latter is approximately 2.71828 and is normally referred to as the natural logarithm. The more precise way of writing the logarithm is to include its base:

$$\log_{10} 1000 = 3 \quad (50)$$

The logarithm in Equation 47 is to the power of Euler's constant e. To find out how many prime numbers there are below or equal to the number n, we can calculate:

$$\frac{n}{\log_e n} \quad (51)$$

The average gap g between primes up to the number n is then:

$$g_n = \frac{n}{\text{number of primes}} \quad (52)$$

We do already know the number of primes from Equation 51, and we can put it into Equation 52:

$$g_n = \frac{n}{\dfrac{n}{\log_e n}} \quad (53)$$

Figure 79: German mathematician Leonhard Euler (1707 - 1783).

We can simplify to:

$$g_n = \frac{n \times \log_e n}{n}$$

$$= \frac{\cancel{n} \times \log_e n}{\cancel{n}} \tag{54}$$

$$= \log_e n$$

The average gap between primes now only depends on the properties of the natural logarithm. Figure 80 shows its plot. The most important property of the natural logarithm for prime numbers is that it will slowly grow to infinity. This means that average gap between primes also increases to infinity, which in turn means that the frequency of prime numbers decreases.

Figure 80: A plot of the natural logarithm $\log_e$.

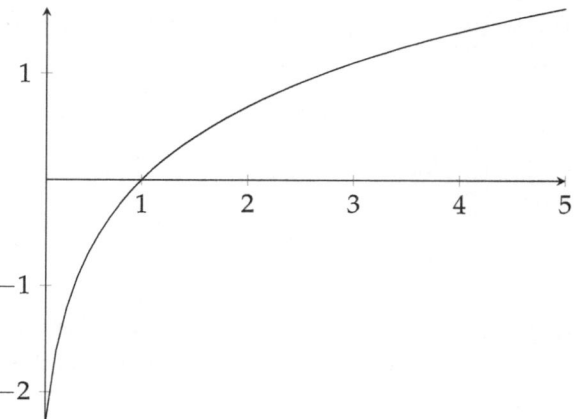

The gap between 11 and 13 is just 2 while the gap between 101 and 107 is already 6 and it grows to 30 for 1 000 003 and 1 000 033. Equation 54 refers to the *average* gap between prime numbers. There will continue to be primes that have just one number between them, so called twin prime numbers. The smallest twin prime numbers are three and five. But they also occur further out, such as 1 000 037 and 1 000 039. Still, on average the gap between them increases, which makes it harder and harder to find them.

Searching for Primes

Modern computers can easily calculate the prime numbers up to 100 000 by using a simple trial and error method (see Listing 23). The main loop in line 18 checks every integer in the search range with the is_prime function. This function starting on line 5 then checks all the possible divisors to see if the remainder of the division, its modulus, is zero. If it is, then we have found a divisor which means that this number is not a prime. This program keeps track of the execution duration which is approximately 75 seconds for a search range of up to 100 000.

You can download this Python program from our repository.

```python
# Simple Python program to find prime numbers in a range
import time
# prime search function
def is_prime(n):
 if n <= 1:
  return False
 for i in range(2, n):
  if n % i == 0:
   return False
 return True
# setup variables
start_time = time.time() # start time
count_primes = 0 # for counting
search_range = 100000 # search up to this value
for n in range(1, search_range): # main loop
 x = is_prime(n)
 count_primes += x
# report result
print("Total prime numbers in range :", count_primes)
stop_time = time.time()
print("Time required :", stop_time - start_time)
```

Listing 23: Simple search for prime numbers.

The problem with this simple search is that it does not scale well. To search up to 10 the program would have to run through the first loop 10 times, which each then has to run through a loop of 10. That makes $10 \times 10 = 100$ steps. More generally, this algorithm requires N^2 iterations. If we are searching for up to $100\,000$ then we will need $100\,000^2 = 10\,000\,000\,000$ steps. We can optimise this approach in several ways. First, we do not need to check all the possible divisors. Let's consider the example of the number 36, which can be factorised as:

$$
\begin{aligned}
36 = & 1 \times 36 \\
& 2 \times 18 \\
& 3 \times 12 \\
& 4 \times 9 \\
& 6 \times 6 \\
& 9 \times 4 \\
& 12 \times 3 \\
& 18 \times 2 \\
& 36 \times 1
\end{aligned}
\tag{55}
$$

Notice that the factors are mirrored. Both, 2×18 and 18×2 occur. This means that we only need to search up to the square root. In case of 36 this would be $\sqrt{36} = 6$. Another optimisation is to exclude all multiples of 2 except 2 itself. Although there is some debate since 2 is the only even prime number. All even numbers cannot be prime numbers. These optimisations can reduce the execution time to 32 seconds, but the scaling problem remains.

The largest prime ever found is a Mersenne Prime (see page 122) which can best be expressed as $2^{136\,279\,841} - 1$. If you want to express this as a normal number, it would have $41\,024\,320$ digits. Luke Durant found it in October 2024 as part of the GIMPS project. If you have 18MB of space on your computer left, then you can download a ZIP file with the number: https://www.mersenne.org/primes/digits/M136279841.zip

Sieve of Eratosthenes

The best approach to finding all the prime numbers up to a given number is an ancient algorithm called the Sieve of Eratosthenes. It was first mentioned in Nicomachus of Gerasa's (ca 60-120) "Introduction to Arithmetic". This algorithm finds prime numbers not by testing if a certain number is a prime, but by removing all non-prime numbers from a list:

1. Create a list of consecutive integers from 2 through n: (2, 3, 4, 5, 6, ... n).

2. Start with p equals 2, which is the smallest prime number.

3. Remove all the multiplies of p, such as $2p = 4, 3p = 6$ from the list. This excludes all the even numbers greater than two.

4. Find the smallest number in the list that is greater than p that has not yet been removed. If there is no such number then stop.

5. Otherwise, let p be this new number and repeat from step 3.

6. When the algorithm stops, all remaining numbers in the list are the primes numbers below n.

Let's see this algorithm in action. We start with a list of integers up to 30:

$$2,3,4,5,6,7,8,9,10,11,12,13,14,15,16,17,18,19,20,21,22,23,24,25,26,27,28,29,30 \tag{56}$$

We start with $p = 2$ and remove all its multiples (step 3):

$$2,3,4,5,6,7,8,9,10,11,12,13,14,15,16,17,18,19,20,21,22,23,24,25,26,27,28,29,30 \tag{57}$$

The next smallest number for p is 3 (step 4). We then remove all its multiples that are not already removed which are 9, 15, 21 and 27.

$$2,3,4,5,6,7,8,9,10,11,12,13,14,15,16,17,18,19,20,21,22,23,24,25,26,27,28,29,30 \tag{58}$$

The next smallest number for p is 5 (step 4). We then remove all its multiples that are not already removed, which is only 25.

$$2,3,4,5,6,7,8,9,10,11,12,13,14,15,16,17,18,19,20,21,22,23,24,25,26,27,28,29,30 \tag{59}$$

The next number to consider would be seven, but all its multiples are already removed. The same holds true for all other numbers not yet removed. Hence the algorithm stops. What remains are the prime numbers up to 30:

$$2,3,5,7,11,13,17,19,23,29 \tag{60}$$

We found them by eliminating all non-prime numbers. In code this would be like Listing 24. Its execution time is only 0.02 seconds.

You can download this Python program from our repository.

```
1   import time
2   def sieve_of_eratosthenes(limit):
3       primes = []
4       is_prime = [True] * (limit + 1)
5       is_prime[0] = is_prime[1] = False
```

```
6
7        for num in range(2, int(limit**0.5) + 1):
8            if is_prime[num]:
9                primes.append(num)
10               for multiple in range(num * num, limit + 1, num):
11                   is_prime[multiple] = False
12
13       for num in range(int(limit**0.5) + 1, limit + 1):
14           if is_prime[num]:
15               primes.append(num)
16
17       return primes
18
19   # Example: Find prime numbers up to 50
20   start_time = time.time()
21   limit = 100000
22   prime_numbers = sieve_of_eratosthenes(limit)
23   stop_time = time.time()
24   print("Prime numbers up to", limit, "are:", prime_numbers)
25   print("Time required:", stop_time - start_time)
```

Listing 24: Sieve Of Eratosthenes program to find prime numbers.

The most important rule in programming is not to reinvent the wheel. SymPy is a Python library for symbolic mathematics[24] that already includes an isPrime() function. It uses different algorithms to test if a number is a prime number based on the size of the number in question. Algorithms used include a search for trivial factors, a bisection search on the sieve, deterministic Miller–Rabin tests and a strong BPSW test.

[24] https://www.sympy.org/en/index
.html

But if you are interested in creative ways to search for primes then there is no way around the Regular Expression. This cryptic computer code .?|(..+?)\\1+ is all it takes to test if a number is prime. To understand how these 13 characters instruct a computer to test if a number is a prime number requires a mere 4782 words[25].

[25] https://illya.sh/the-codeument
ary-blog/regular-expression-check
-if-number-is-prime/

Time and Space Complexity

For any algorithm to run, it is worthwhile to consider its efficiency for its worst-case scenario. We start by defining its input N. In our example in Listing 23 we looked at numbers up to 100 000. It executes in N^2 steps. In computer science and maths, we use the Big O notation to describe the efficiency of an algorithm; more specifically we call it the time complexity. In this case, we can describe the algorithm's complexity as $O(N^2)$. The Sieve of Eratosthenes algorithm has a complexity of $O(N \log(\log N))$. Paul Pritchard optimised the sieve approach in 1979. His "Sieve of Pritchard" has the Big O notation of $O(N/\log \log N)$.

The algorithm to search for the Recaman sequence on page 86 has a pretty high complexity of $O(N^2)$ while the search for Fibbonaci numbers on page 80 uses only $O(N)$. Figure 81 shows typical examples of time complexity. Besides estimating the algorithm's execution time, it is also necessary to consider how much memory it will require. This is referred to as the algorithm's space complexity since every transistor that stores a value occupies space. There are only approximately 10^{80} atoms in the observable universe, which sets a natural boundary on all space considerations. Even if an algorithm can complete a task in a realistic time, if the result requires more memory than is available, the result cannot be stored.

Figure 81: Examples of Big O complexities.

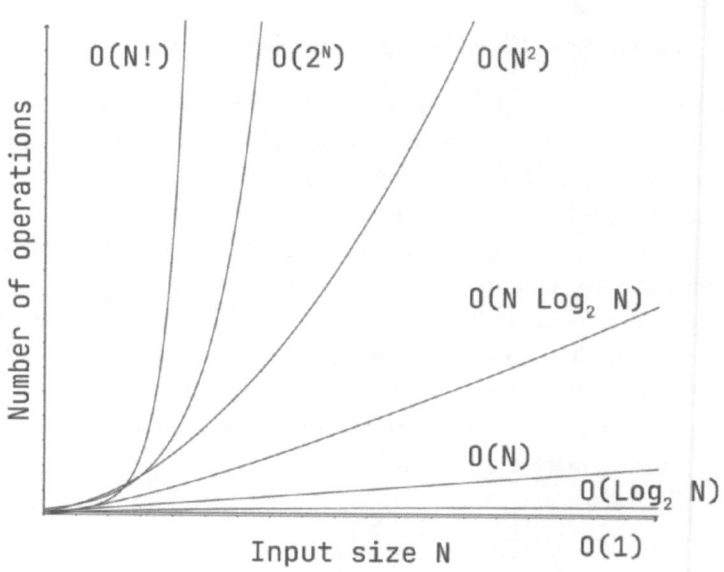

Bonus

A famous example of the danger of underestimating the space complexity involves the invention of chess. According to legend, Grand Vizier Sissa ben Dahir was asked by the Indian King Shirham what reward he wanted for inventing the game of chess. Sissa replied that on the first day, he would want one grain of wheat to be placed on the first square of the chessboard. On the second day, he wanted two grains of wheat placed on the second square, on the third day four grains on the third square and so forth (see Figure below for the first four squares). Shirham was amazed by this seemingly cheap request.

The time complexity is simply $O(64)$ since a chessboard has $8 \times 8 = 64$ squares. The catch is with the space complexity which is $O(2^N)$. While $2^{64} - 1$ looks harmless, it does add up to an astonishing $18\,446\,744\,073\,709\,551\,615$, which is $1\,199\,000\,000\,000$ metric tons. This is far more than the annual worldwide grain production.

A more recent example of the power of exponential growth is Russia's penalty fees towards YouTube for blocking Russian TV channels from uploading content. The exact details of the court ruling are unclear. The news widely reported on a $20 decillion penalty fee. That is a 2 followed by 34 zeroes. Supposedly, the original penalty fee was set to $100\,000$ Rubles per day. The daily fee would double every week. The exact starting date is unclear, but Google's Russian legal entity filed for bankruptcy in 2022. So let's assume that the fees racked up for two years. This would bring us to $14\,197\,686\,722\,556\,200\,000\,000\,000\,000\,000\,000$. This number has 38 digits. Or to express this in Google's terms:

Goooooooooooooooooooooooooooooooooogle.

No matter if the fee has a total of 35 or 38 digits, it is more than all of the money in the world. The Russian judges exposed their mathematical incompetence by using an exponential growth function for the penalty fees.

Swimming Prime Numbers

It is faster to search for numbers that are not prime numbers than for those that are. When swimming the prime numbers, think not about what you swim, but what you do not swim.

Figure 82: Swim the Prime Numbers swimming program. You can download this swimming program as a PDF from our repository.

Warm up	
200 Any Easy	1
Prime Numbers	
1 × **100** FR ⏱0:15	2
4 × **100** IM ⏱0:15	3
6 × **100** FR ⏱0:15	4
8 × **100** IM Overlap ⏱0:15 *Overlap 50s*	5
9 × **100** FR ⏱0:15	6
Warm down	
100 Any Easy	7

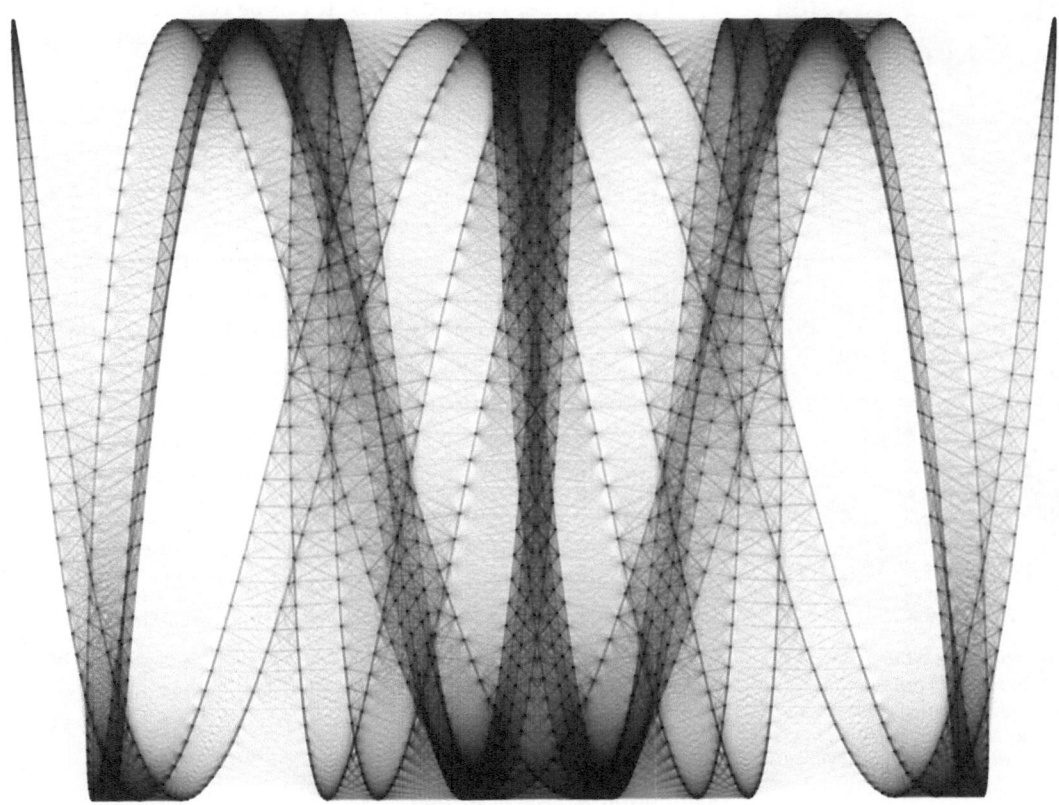

Absolute Primes

Now that we know what primes are and how to find them, we can go one step further. We are now looking for primes that are also circular. This prime's digits could be rotated and still result in a prime. The easiest example would be 11. You can rotate the two ones around and would still get, well, 11. 111 is not a prime since it is $111 = 3 \times 37$. The next circular prime that is made up of only 1s has 19 digits. While there can be an infinite number of these repunit primes, they still remain rather boring and will be excluded from further consideration in this section. When we want to more systematically search for circular primes, we can already exclude any number that has a digit of 2 in it. If we rotate around this number, eventually the 2 will move to the bac, which will result in an even number. Let's consider 323. It can be rotated to 233 and 332. The latter is an even number and therefore not prime. The same holds true for the digits 0,4,6 and 8. Five is also not a good digit, since any number ending in five is divisible by five. This leaves us with 1,3,7 and 9.

The lowest two digit circular prime is 13 since 31 is also a prime. The same holds true for 17 and 71. There are a few more two, three, four and five digit circular primes: 37, 79, 113, 197, 199, 337, 1 193, 3 779, 11 939, 19 937. The largest known six digit circular primes are 193 939 and 199 933. The important thing here is that it is a finite set.

So far, we constrained the numbers to circular permutations such as 197,971 and 719. But what happens if we allow all the digits to be moved to any other place in any order? This means that we would allow 179, 917 and 791 from the previous example. If all the possible permutations are also prime numbers, then we have found an absolute prime. Unfortunately, 791 is not a prime number since $791 = 7 \times 113$.

We do exclude the single and double digit primes as the only permutations possible with these are circular. Since all absolute primes must be also a circular prime, we can dramatically reduce the search space. The only three known absolute primes are 113, 199 and 337. Since two digits in each of these absolute primes are identical, all permutations of them are also just circular: 113, 131, 311; 199, 919, 991; 337, 373, 733.

Swimming Absolute Primes

We can swim the absolute primes by using the digits as lap counters (see Figure 84). Feel free to add further variations to the programs, such as intensity levels, as you desire.

This sequence is registered with the On-Line Encyclopedia of Integer Sequences as A068652 (https://oeis.org/A0686 52)

Figure 83: When asking AI to generate a visualisation of absolute prime numbers a stereotypical library with a chalk board emerges. The formulas on the board are gibberish. Even more interesting is the question of how anybody could sit in the left armchair.

Warm up			
	1 laps FL		1
125 as	**1** laps BK		2
	3 laps FR		3
	1 laps BK		4
125 as	**3** laps FR		5
	1 laps FL		6
	3 laps FR		7
125 as	**1** laps FL		8
	1 laps BK		9
1 laps Any Easy			10

First set			
	1 laps FL		11
475 as	**9** laps BK		12
	9 laps FR		13
	9 laps BK		14
475 as	**9** laps FR		15
	1 laps FL		16
	9 laps FR		17
475 as	**1** laps FL		18
	9 laps BK		19
1 laps Any Easy			20

Second set			
	3 laps BR		21
325 as	**3** laps BK		22
	7 laps FR		23
	3 laps BK		24
325 as	**7** laps FR		25
	3 laps BR		26
	7 laps FR		27
325 as	**3** laps BR		28
	3 laps BK		29
1 laps Any Easy			30

Figure 84: Swim the Absolute Prime Numbers swimming program. You can download this swimming program as a PDF from our repository.

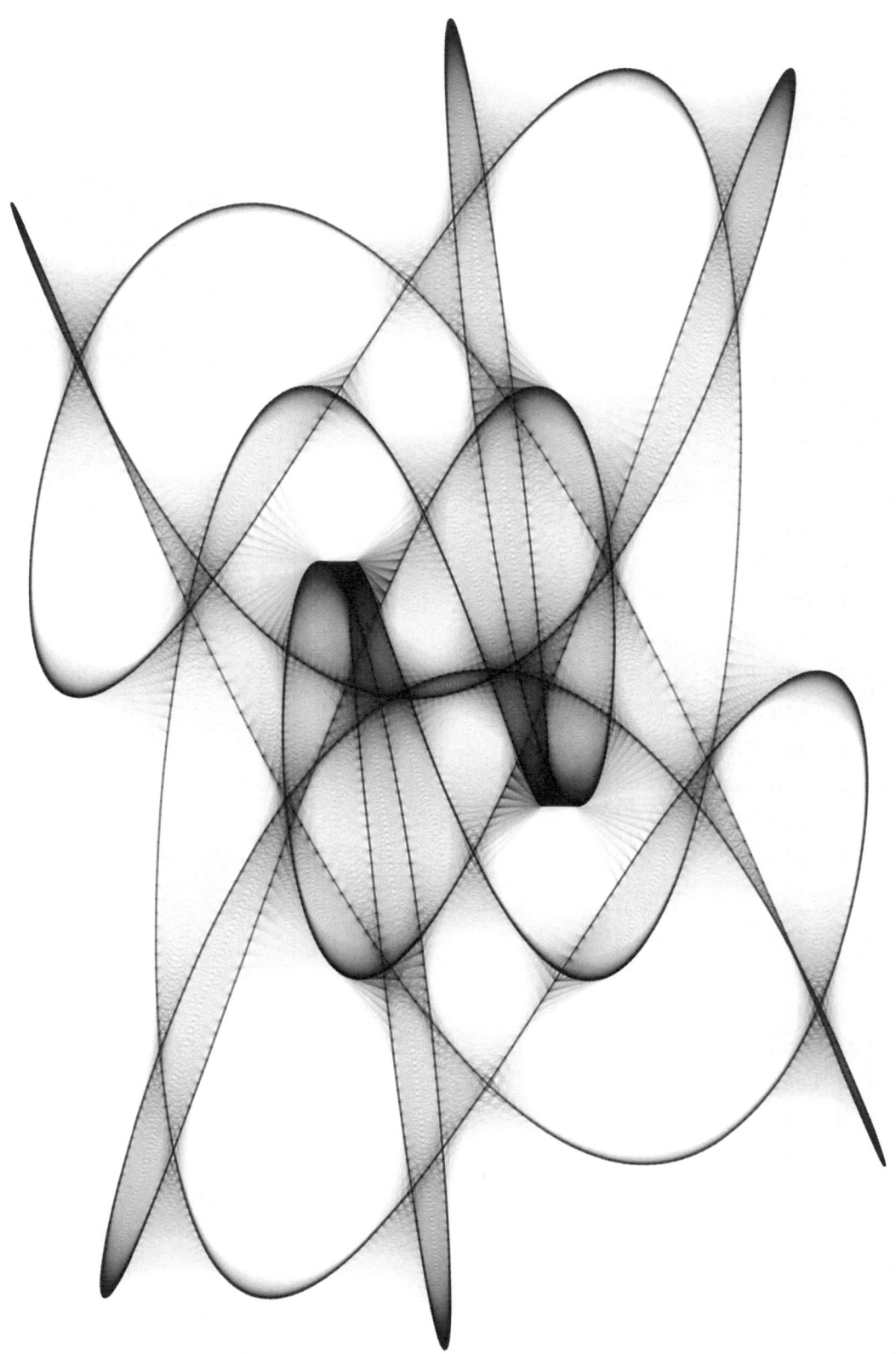

Highly Composite Numbers

Prime numbers are only divisible by one and themselves. This means that they have the minimum factors. Seven, for example, is only divisible by one and seven:

$$7 = 1 \times 7 \tag{61}$$

We might ask, if prime numbers are numbers with the least number of factors, what are the numbers with the most factors? Let's try another example. The number 12 is divisible by a total of six factors:

$$\begin{aligned} 12 &= 1 \times 12 \\ &= 2 \times 6 \\ &= 3 \times 4 \end{aligned} \tag{62}$$

This might look like a lot, but the number of possible factors does of course increase the larger the number gets. The Greek Philosopher Plato (see Figure 85) considered the number 5040 as being a great example of a number with many divisors. In his book "The Laws", he wrote that 5040 would be a great number of citizens for a city (Schofield and Griffith, 2016). He argued that 5040 is divisible by 60 factors which is convenient to divide land and many other resources amongst the citizens. Moreover, it is divisible by the numbers of 1-10, and it is divisible by more factors than any number before it.

Numbers that have such a high number of factors are called Highly Composite Numbers (HCN). They are the opposite of prime numbers in this sense. The mathematician Srinivasa Ramanujan investigated the properties of these numbers in more detail. He defined a Highly Composite Number as a number that has more factors than the number before it.

Let's start at the beginning. The number 1 is only divisible by one number. This is more than the number 0, and hence it qualifies as an HCN. The number 2 is a prime number, since it is divisible only by 1 and itself. It has therefore two factors. This is more than any number before it, so it is also an HCN. We mark HCNs in Table 7 in the bold typeface. All prime numbers are only divisible by 2 factors. This insight is useful for the following sequence. The number 3 is another prime, but since 2 came before it, it is not an HCN. The number 4 is the first with three factors (1, 2, 4), and 6 is the first with four factors (1, 2, 3, 4). We then have to count up to 12 to find the first number that has more factors than 6. It has the six factors shown in Equation 62.

Figure 85: Plato (427-348 BC) was a Greek philosopher. Many consider him the founder of Western philosophy (source: Marie-Lan Nguyen).

This sequence is registered with the On-Line Encyclopedia of Integer Sequences as A002182 (https://oeis.org/A0021 82)

n	1	2	3	4	5	6	7	8	9	10	11	12
$d(n)$	**1**	**2**	2	**3**	2	**4**	2	4	3	4	2	**6**

Table 7: Highly Composite Numbers.

When we write out the sequence of HCN we get:

$$1, 2, 4, 6, 12, 24, 36, 48, 60, 120, 180, 240, 360, 720, 840, 1260, 1680, 2520, 5040 \tag{63}$$

To better understand this sequence, we have to take a slight detour to the Fundamental Theorem of Arithmetic. It can be traced back to Euclid (see Figure 89) but the first formal proof was done by the German mathematician Carl Friedrich Gauss (see Figure 86). It states that:

Fundamental Theorem of Arithmetic 1 *Every integer greater than 1 can be represented uniquely as a product of prime numbers, up to the order of the factors.*

More formally it can be written as:

$$n = p_1^{a_1} \times p_2^{a_2} \times p_3^{a_3} \times ... \times p_k^{a_k} \tag{64}$$

Where p are prime numbers and a are exponents. This might look somewhat complicated, but it will become much clearer when considering an example. The number 30, for example, can be represented as:

$$550 = 2^1 \times 5^2 \times 11^1 \tag{65}$$

Prime numbers are the building blocks of which all other numbers can be made. This brings us one step closer to finding the answer to how many factors a number has. Table 7 listed a couple of examples, but we want a more general solution.

$$d(n) = (a_1 + 1) \times (a_2 + 1) \times (a_3 + 1) \times ... \times (a_k + 1) \tag{66}$$

When we want to know the total number of factors for , then we only need to use the exponents a from Equation 65 in Equation 66:

$$d(550) = (1 + 1) \times (2 + 1) \times (1 + 1) = 12 \tag{67}$$

The factors are 1, 2, 5, 10, 11, 22, 25, 50, 55, 110, 275, and 550. When we want to calculate the number of factors for 5040, we first have to again find its prime factors. They are:

$$5040 = 2^4 \times 3^2 \times 5^1 \times 7^1 \tag{68}$$

It follows that:

$$d(5040) = (4 + 1) \times (2 + 1) \times (1 + 1) \times (1 + 1) = 60 \tag{69}$$

When Srinivasa Ramanujan (see Figure 87) studied HCNs, he discovered three properties that they need to have. The first one is that the prime factors need to be in consecutive order. Notice that the prime factors for 5040 (shown in Equation 68) are 2,3,5 and 7 while the prime factors of 550 (shown in Equation 65) are not: 2,5, and 11. The prime factor 7 is missing in the sequence. 550 is therefore not an HCN because a smaller number with the same number of factors exists. We will keep the same number of exponents as for 550 but use consecutive prime factors:

$$90 = 2^1 \times 3^2 \times 5^2$$
$$d(90) = (1 + 1) \times (2 + 1) \times (1 + 1) = 12 \tag{70}$$

The number 90 has the same number of factors as 550, but it is smaller. 550 is therefore not an HCN.

The second property is that the exponents need to be in a weakly descending order. Weakly descending means that value of each term is

Figure 86: Carl Friedrich Gauss (1777-1855) was a German mathematician. He is one of the most important mathematician of the modern world (Painted by Christian Albrecht Jensen).

Figure 87: Srinivasa Ramanujan (1887–1920) was an Indian mathematician.

less than or equal to the previous term. If we look at 5040, then we notice that the exponents are 4, 2, 1 and 1. The exponents for 90 are 2, 1 and 1. They do not weakly descend, and therefore, 90 is also not a HCN. There is a better choice. If we keep the sequence of the prime factors the same but swap the exponents into a descending order, then we get:

$$2^2 \times 3^1 \times 5^1 = 60 \tag{71}$$

The number 60 has the same number of prime factors as 90 but is smaller. The third property is that the last exponent has to be 1. The number 60 has this property and hence is an HCN. There are two exceptions to this third property. The number 4 can be prime factorised as $4 = 2^2$, and 36 as $36 = 2^2 \times 2^3$. Both are HCNs.

Searching for Highly Composite Numbers

We can write a simple Python program to search for Highly Composite Numbers. We will use the simple but inefficient algorithm to find the factors of a number already discussed in Listing 23. We only have to extend this algorithm with keeping track of the maximum number of factors already found (line 16).

You can download this Python program from our repository.

```python
def count_divisors(n):
    #Returns the number of divisors of n
    count = 0
    for i in range(1, int(n**0.5) + 1):
        if n % i == 0:
            count += 2 if i != n // i else 1
    return count
def find_highly_composite_numbers(limit):
    #Finds all Highly Composite Numbers up to a given limit
    highly_composite_numbers = []
    max_divisors = 0
    for num in range(1, limit + 1):
        divisors = count_divisors(num)
        if divisors > max_divisors:
            highly_composite_numbers.append(num)
            max_divisors = divisors
    return highly_composite_numbers
# Example usage
limit = 100
hcn_list = find_highly_composite_numbers(limit)
print(f"Highly Composite Numbers up to {limit}: {hcn_list}")
```

Listing 25: Not primes.

Swimming Highly Composite Numbers

Highly Composite Numbers are ideal for dividing up a swimming program. The HCNs in Equation 63 give you many options. You can interpret the numbers as the number of laps. If we take 120 laps in a 25 meter pool, then we get to 3000 meters. Its factors are:

$$
\begin{aligned}
120 &= 1 \times 120 \\
&= 2 \times 60 \\
&= 3 \times 40 \\
&= 4 \times 30 \\
&= 5 \times 24 \\
&= 6 \times 20 \\
&= 8 \times 15 \\
&= 10 \times 12
\end{aligned}
$$

(72)

You can now choose to break up your program into any of these even parts. For this example, we choose 5×24 laps which is 5×600 meters.

Figure 88: Highly Composite Numbers swimming program. You can download this swimming program as a PDF from our repository.

120 laps			
600 as	**100** FR		1
	100 D Any Any		2
	100 Not FR		3
6 × **100**	IM @_1:45		4
4 × **150**	D IM Order Any ↻0:15		5
3 × **200**	FR ↻0:15 Pads		6
2 × **300**	FR ↻0:15 Pads		7

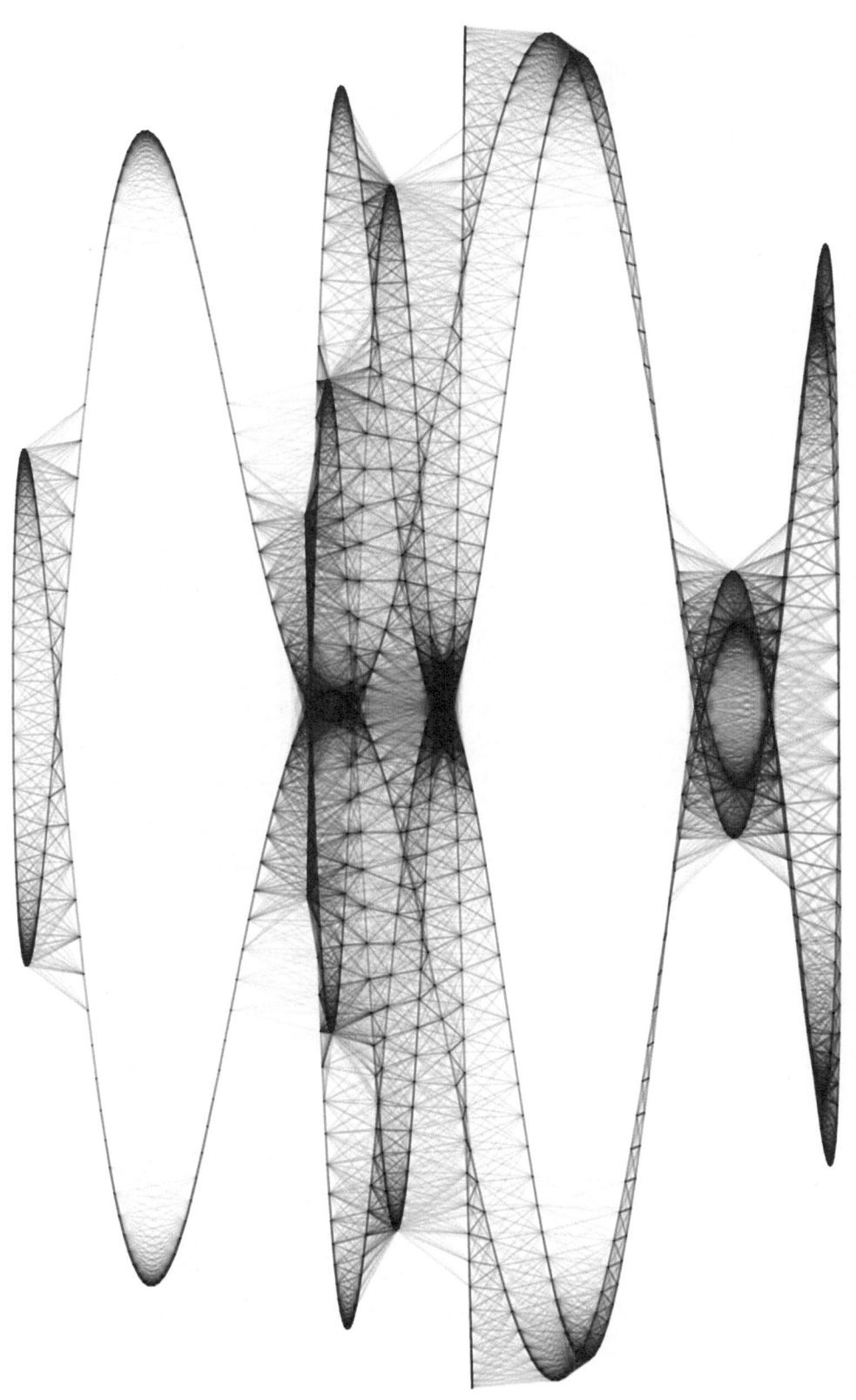

Perfect Number

Many swimmers seek the perfect time, drill, distance or stroke. Numbers can also be perfect. A Perfect Number is a positive integer that is equal to the sum of its positive divisors, excluding the number itself. Let's consider an example. The number six can be factorised as:

$$\begin{aligned} 6 &= 1 \times 6 \\ &= 2 \times 3 \end{aligned} \tag{73}$$

Its factors are therefore 1, 2, 3, and 6. We exclude the number itself and are left with:

$$1 + 2 + 3 = 6 \tag{74}$$

The number 6 is, therefore, considered perfect. The next perfect number is 28:

$$\begin{aligned} 28 &= 1 \times 28 \\ &= 2 \times 14 \\ &= 4 \times 7 \\ 28 &= 1 + 2 + 4 + 7 + 14 \end{aligned} \tag{75}$$

The ancient Greeks only found the first four perfect numbers. In order to find more, it was necessary to find the pattern that predicts their occurrences. The numbers can also be expressed as:

$$\begin{aligned} 6 &= 1 + 2 + 3 \\ 28 &= 1 + 2 + 3 + 4 + 5 + 6 + 7 \\ 496 &= 1 + 2 + 3 + 4 + 5 + 6 + 7 + \ldots + 30 + 31 \\ 8128 &= 1 + 2 + 3 + 4 + 5 + 6 + 7 + \ldots + 126 + 127 \end{aligned} \tag{76}$$

There are even more patterns. They might not all be useful, but they are fascinating in themselves. All but 6 can also be expressed as:

$$\begin{aligned} 28 &= 1^3 + 3^3 \\ 496 &= 1^3 + 3^3 + 5^3 + 7^3 \\ 8128 &= 1^3 + 3^3 + 5^3 + 7^3 + \ldots + 15^3 \end{aligned} \tag{77}$$

The story of interesting patterns continues. If we write the perfect numbers in binary (see the section Binary on page 63), we get a sequence of ones followed by a sequence of zeros:

$$\begin{aligned} 6_{10} &= 110_2 \\ 28_{10} &= 11100_2 \\ 496_{10} &= 111110000_2 \\ 8128_{10} &= 1111111000000_2 \end{aligned} \tag{78}$$

This means that these numbers are consecutive of 2:

$$\begin{aligned} 6 &= 2^2 + 2^1 \\ 28 &= 2^4 + 2^3 + 2^2 + 2^1 \\ 496 &= 2^8 + 2^7 + 2^6 + 2^5 + 2^4 \\ 8128 &= 2^{12} + 2^{11} + 2^{10} + 2^9 + 2^8 + 2^7 + 2^6 \end{aligned} \tag{79}$$

Finding Perfect Numbers

Euclid (see Figure 89) was the first to find a pattern for predicting the occurrence of perfect numbers. We have to start with the sequence of the multiple of 2:

$$1, 2, 4, 8, 16, 32, 64... \tag{80}$$

We take the first two numbers and add them:

$$1 + 2 = 3 \tag{81}$$

If the sum is a prime number, then we multiply it with the last number in the sequence:

$$2 \times 3 = 6 \tag{82}$$

We continue in the sequence:

$$1 + 2 + 4 = 7 \tag{83}$$

7 is a prime number, so we multiply it with the last number in the sequence:

$$4 \times 7 = 28 \tag{84}$$

Let's continue. The next iteration is:

$$1 + 2 + 4 + 8 = 15 \tag{85}$$

15 is not a prime number, so we continue:

$$1 + 2 + 4 + 8 + 16 = 31 \tag{86}$$

31 is a prime number, so we calculate:

$$16 \times 31 = 496 \tag{87}$$

There is a more general way to express this pattern. We can express them as:

$$\begin{aligned} 6 &= (1 + 2) \times 2^1 \\ 28 &= (1 + 2 + 4) \times 2^2 \\ 496 &= (1 + 2 + 4 + 8 + 16) \times 2^4 \end{aligned} \tag{88}$$

Where the first term is a prime number. We can express this in an even more elegant way. Let's focus on the first term (i.e. $1 + 2 + 4$). We can write this more generally as:

$$2^0 + 2^1 + 2^2 + ... + 2^{n-2} + 2^{n-1} = T \tag{89}$$

We do not know what n is, so we equate it for now to T. If we multiply both sides of the equation with 2 we get:

$$2^1 + 2^2 + 2^3 + ... + 2^{n-1} + 2^n = 2T \tag{90}$$

If we subtract Equation 89 from Equation 90, we notice that most terms cancel each other out:

$$\begin{aligned} \cancel{2^1} + \cancel{2^2} + \cancel{2^3} \quad + ... + \cancel{2^{n-1}} \quad + 2^n &= 2\,T \\ 2^0 + \cancel{2^1} \quad + \cancel{2^2} + ... \quad + \cancel{2^{n-2}} + \cancel{2^{n-1}} \qquad\qquad &= \quad T \\ \hline -1 + \qquad\qquad\qquad\qquad\qquad\qquad 2^n &= \quad T \\ 2^n - 1 &= \quad T \end{aligned} \tag{91}$$

Figure 89: Euclid (around 300BC) was a Greek mathematician, and he is considered by many the "father of geometry". This painting from around 1630 is by Jusepe de Ribera.

We now replace the first term in Equation 88 with T where n will equal the number of elements in the sequence:

$$6 = (2^2 - 1) \times 2^1$$
$$28 = (2^3 - 1) \times 2^2 \qquad\qquad (92)$$
$$496 = (2^5 - 1) \times 2^4$$

We notice that the exponent in the first term is always one more than the exponent in the second term. If we call the exponent p, the formula for perfect numbers is:

$$(2^p - 1) \times 2^{p-1} \qquad\qquad (93)$$

When the first term $(2^p - 1)$ is a prime number. The key to finding perfect numbers was to find values for p so that $(2^p - 1)$ is a prime number. The first value for p for which this holds true is 2:

$$(2^2 - 1) \times 2^{2-1}$$
$$3 \times 2 \qquad = 6 \qquad\qquad (94)$$

The next value for p is 3:

$$(2^3 - 1) \times 2^{3-1}$$
$$7 \times 4 \qquad = 28 \qquad\qquad (95)$$

In 1644 Marin Mersenne published his book "Cogitata Physica Mathematica" (see Figure 90) in which he listed the first 11 values for p for which he claimed that $(2^p - 1)$ is a prime number. He was correct for the first eight values but made some mistakes afterwards. Prime numbers that comply with this form are called "Mersenne Primes".

The search for perfect numbers is, therefore, linked to tests for Mersenne Prime Numbers. In the section Prime Numbers we learned that searching for prime numbers is a struggle even for modern computers.

The first eight Perfect Numbers are listed in Table 8. There are currently just 51 Perfect Numbers found. These correspond to the 51 Mersenne Primes found. It is assumed that there are infinitely many, but finding them is a difficult problem. The largest one, which was found in 2018, contains more than 23 million digits and is best expressed by the formula $2^{82\,589\,932} \times (2^{82\,589\,933} - 1)$.

Figure 90: Cogitata Physica Mathematica (1644).

Table 8: The first eight Perfect Numbers. This sequence is registered with the On-Line Encyclopedia of Integer Sequences as A000396 (https://oeis.org/A000396)

#	Perfect Number
1	6
2	28
3	496
4	8 128
5	33 550 336
6	8 589 869 056
7	137 438 691 328
8	2 305 843 008 139 952 128

Testing a Number for Perfection

We can write a little Python program (see Listing 26) that tests if a number is perfect. We use a simple, but inefficient, exhaustive search for finding divisors using a loop in lines 3-5. This is similar to our first approach to finding prime numbers (see Listing 23). We then test if the sum of the divisors is equivalent to the number itself (line 7).

```python
def is_perfect_number(num):
    divisors = []
    for i in range(1, num):
        if num % i == 0:
            divisors.append(i)

    if sum(divisors) == num:
        return True
    else:
        return False

# Test the function with a specific number
number = 28  # Example number
if is_perfect_number(number):
    print(number, "is a perfect number.")
else:
    print(number, "is not a perfect number.")
```

Listing 26: Testing if a number is perfect.

With this program we can find any number of perfect numbers, but since we already know how rare they are (see Table 8), even an optimised algorithm would be unable to find a new one in a reasonable time on regular computers.

The 8th perfect number was found by Leonhard Euler (see Figure 79) in 1772 using trial division. It was the first new perfect number discovered for 125 years.

Swimming Perfect Numbers

We can use the digits of the 8th perfect number in blocks of three and round them up to the closest 50 (see Table 9).

x	$50 \times \lceil \frac{x}{50} + \frac{1}{2} \rceil$
2	50
305	350
843	850
008	50
139	150
952	1000
128	150

Table 9: Rounding the digits of the 8th perfect number in blocks of three to the nearest 50.

Figure 91: Swim the 8th Perfect Number swimming program. You can download this swimming program as a PDF from our repository.

Warm up		
400 Any Easy		1
Perfect set		
50 as	**50** Any Easy	2
	50 FR b7	3
350 as	**100** FR b5	4
	200 FR b3	5
	100 IM Race Pace	6
	150 FR Easy	7
	100 IM Race Pace	8
850 as	**150** FR Easy	9
	100 IM Race Pace	10
	150 FR Easy	11
	100 IM Race Pace	12
50 as	**50** Any Easy	13
150 as	**150** K Any Fins	14
	100 Nr 4	15
1000 as	**200** Nr 3	16
	300 Nr 2	17
	400 Nr 1	18
150 as	**150** K Any	19

Bonus

Table 10 shows the factors of the 8th perfect number. It is amazing that Euler was able to calculate that $(2^{31} - 1) \times 2^{30-1}$ is a perfect number. And in case you wonder how to pronounce this number to your fellow swimmers, here it is in words: two quintillion three hundred five quadrillion eight hundred forty-three trillion eight billion one hundred thirty-nine million nine hundred fifty-two thousand one hundred twenty-eight.

1	2 147 483 647
2	4 294 967 294
4	8 589 934 588
8	17 179 869 176
16	34 359 738 352
32	68 719 476 704
64	137 438 953 408
128	274 877 906 816
256	549 755 813 632
512	1 099 511 627 264
1 024	2 199 023 254 528
2 048	4 398 046 509 056
4 096	8 796 093 018 112
8 192	17 592 186 036 224
16 384	35 184 372 072 448
32 768	70 368 744 144 896
65 536	140 737 488 289 792
131 072	281 474 976 579 584
262 144	562 949 953 159 168
524 288	1 125 899 906 318 330
1 048 576	2 251 799 812 636 670
2 097 152	4 503 599 625 273 340
4 194 304	9 007 199 250 546 680
8 388 608	18 014 398 501 093 300
16 777 216	36 028 797 002 186 700
33 554 432	72 057 594 004 373 500
67 108 864	144 115 188 008 747 000
134 217 728	288 230 376 017 494 000
268 435 456	576 460 752 034 988 000
536 870 912	1 152 921 504 069 970 000
1 073 741 824	2 305 843 008 139 952 128

Table 10: The factors of the 8th perfect number.

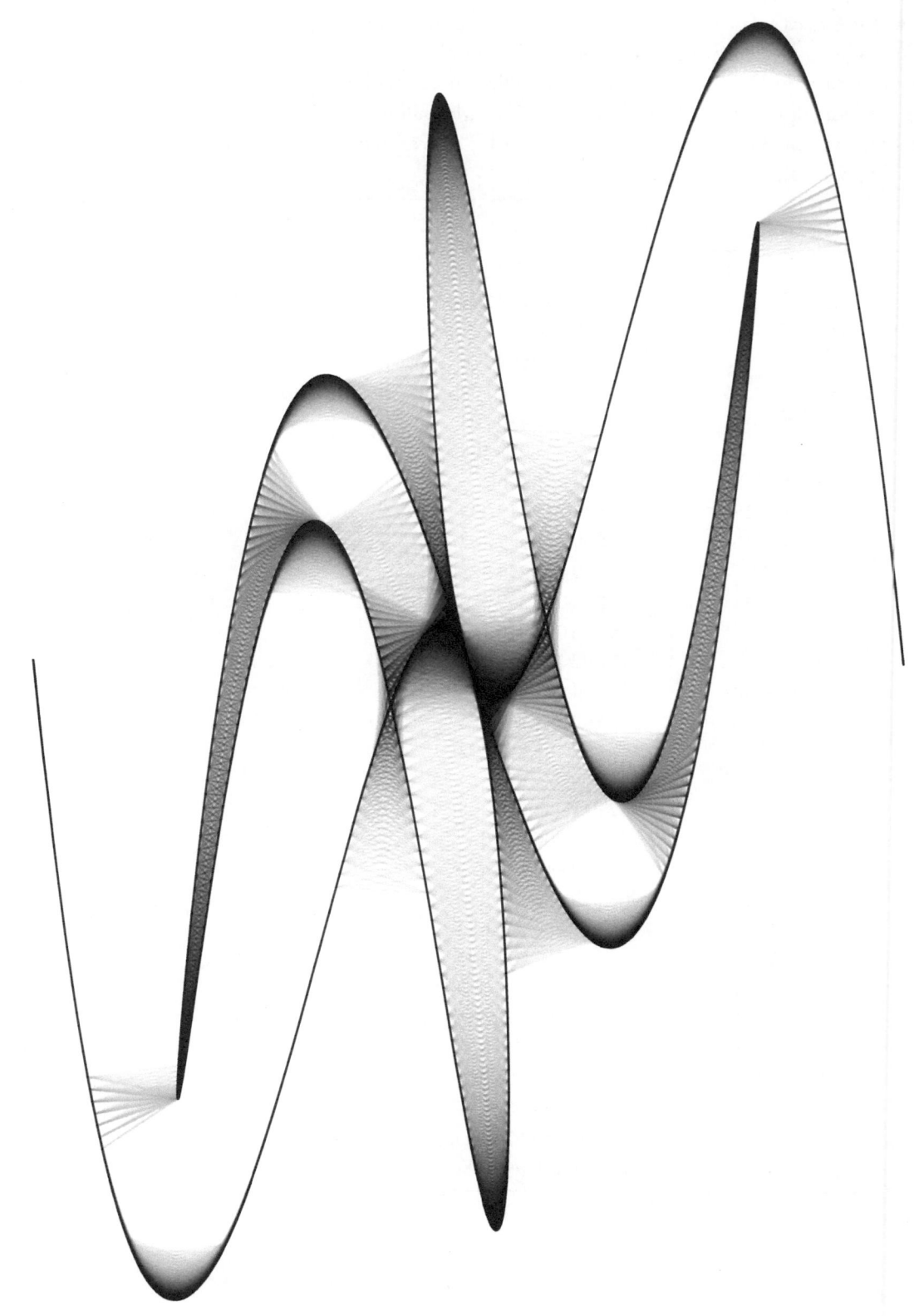

Aliquot

Only a few numbers are perfect. The sum of their proper factors often falls short. If we consider the case of the number 8:

$$8 = 1 \times 8$$
$$= 2 \times 4 \tag{96}$$

When we add up the factors, excluding itself, we get:

$$1 + 2 + 4 = 7 \tag{97}$$

The number 8 is therefore considered a *deficient* number. Most numbers are deficient. A number where the factors add up to more than itself is called an *abundant* number. 24 is a good example:

$$24 = 1 \times 24$$
$$= 2 \times 12$$
$$= 3 \times 8$$
$$= 4 \times 6 \tag{98}$$

When we add up the factors, excluding itself, we get:

$$1 + 2 + 3 + 4 + 6 + 8 + 12 = 36 \tag{99}$$

We can now create a sequence by repeating this step. The factors of 36 add up to 55, which in turn adds up to 17. This is a prime number, and the last step ends therefore in 1. It took three steps before this sequence ended.

This sequence is registered with the On-Line Encyclopedia of Integer Sequences as A001065 (https://oeis.org/A0010 65)

Perfect numbers, of course, continue endlessly, since the sum of their proper factors is themselves. Numbers where the aliquot sequence ends in a perfect number are called *aspiring* numbers. The steps for 95, for example, are: 25 and 6. The latter being of course a perfect number.

But there are other interesting cases. The sequence can also loop. Let us consider the case of 220. Its factors add up to 284:

$$1 + 2 + 4 + 5 + 10 + 11 + 20 + 22 + 44 + 55 + 110 = 284 \tag{100}$$

The factors of 284 add up to 220:

$$1 + 2 + 4 + 71 + 142 = 220 \tag{101}$$

Pairs like this are called *amicable* numbers. Their loop length is two. There are no known loop lengths of order 3. An example of a number whose aliquot sequence is of the order of four is 1 264 460. Its sequence is: 1 547 860, 1 727 636, 1 305 184 and then back to 1 264 460.

The sequence of untouchable numbers is registered with the On-Line Encyclopedia of Integer Sequences as A005114 (https://oeis.org/A005114)

Possibly sadly, depending on your capacity to have empathy for numbers, there are also "untouchable" numbers. These are numbers that are never part of any aliquot sequence. Here are the first six untouchable numbers:

$$2, 5, 52, 88, 96, 120, \ldots \tag{102}$$

Amongst the untouchable numbers, five is currently the only uneven number, making in particularly lonely.

There is one last case of interest. What if we cannot determine if the sequence will ever end in either a prime number, a perfect number, or a set of amicable numbers? This is an open question in mathematics and is often referred to as the Catalan–Dickson conjecture.

You may ask how is it possible that we do not know if a certain number will converge. The answer is that some numbers are particularly abundant. The first 30 steps of the aliquot sequence of the number 276 can only be plotted on a logarithmic scale (see Figure 92).

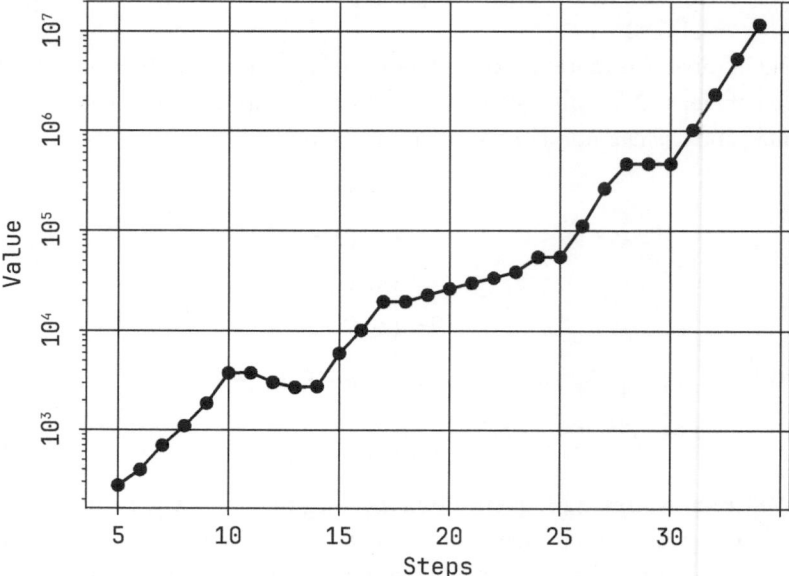

Figure 92: The aliquot sequence of 276 on a logarithmic scale.

276 is the lowest number for which we do not know if it will converge. All attempts to calculate the sequence so far have not converged. Other numbers for which we do not know if they converge are: 552, 564, 660, and 966.

The problem in calculating the aliquot sequence is of course the necessity to find the factors. We already encountered this problem in the section Searching for Primes on page 104. Factorisation is a computationally expensive endeavour.

Swimming the Aliquot Sequence

It is time to dive into the unknown. We can swim the lowest unknown aliquot sequence of 276 as:

Warm up	
200 Any Easy	1
Aliquot sequence	
2 × **200** IM ⏱0:20 80%	2
7 × **200** FR ⏱0:30 70%	3
6 × **200** Nr 1 ⏱0:40 60%	4
Warm down	
200 Any Easy	5

Figure 93: Swim the smallest unknown aliquot sequence of the number 276. You can download this swimming program as a PDF from our repository.

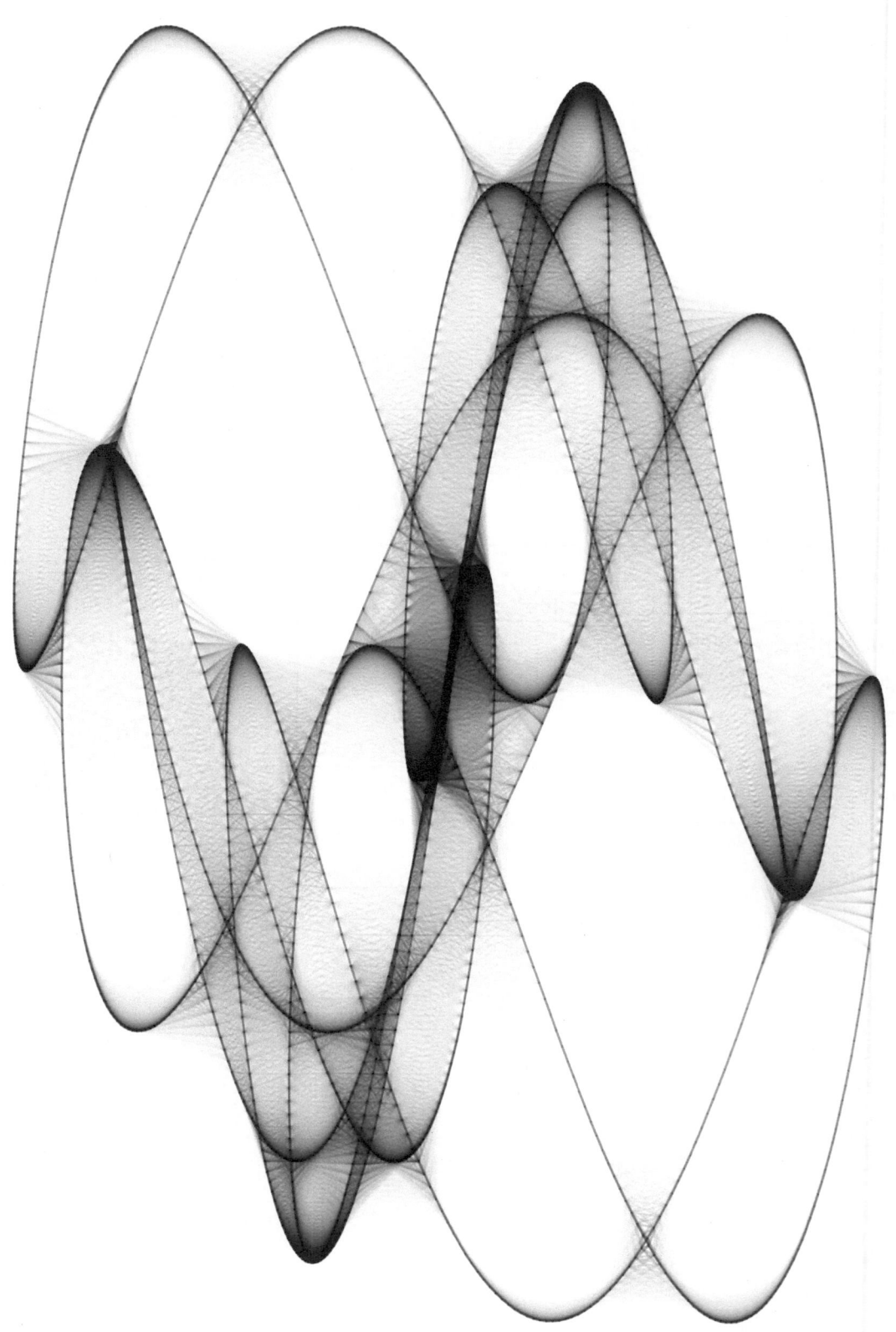

Magic Square

The tiles in a pool can become magical. Let's consider a 3×3 grid of square tiles. We can assign each tile a number from 1-9 so that each row and column adds up to 15 (see Figure 94). Furthermore, the two diagonals also add up to 15.

(i)

2	7	6
9	5	1
4	3	8

(ii)
$$
\begin{array}{ccccccc}
 & & & & & & 15 \\
 & & & & & & = \\
2 & + & 7 & + & 6 & = & 15 \\
+ & & + & & + & & + \\
9 & + & 5 & + & 1 & = & 15 \\
+ & & + & & + & & + \\
4 & + & 3 & + & 8 & = & 15 \\
= & & = & & = & & = \\
15 & & 15 & & 15 & & 15
\end{array}
$$

Figure 94: A 3×3 magic square.

This magic square first appeared in China in the first century A. D. The Ta Tai Li-Chi showed a magic square, but it was based on even older sources. In medieval times it was referred to as the Lo Shu (document of the Lo River), based on a legend involving a King and a turtle.

Figure 95: Melencolia I by Albrecht Dürer (1514).

Arguably the first European magical square can be found in the engraving "Melencolia I" by Albrecht Dürer (see Figure 95). It can be found in the top right square of the print. The middle two numbers of the last row show the year of the engraving: 1514. All rows, columns and diagonals add up to the number 34 (see Figure 96i).

Figure 96: Dürer's 4 × 4 magic square.

(i)

16	3	2	13
5	10	11	8
9	6	7	12
4	15	14	1

(ii)

16	3	2	13
5	10	11	8
9	6	7	12
4	15	14	1

What makes this magic square even more magical is that the four quadrants and the middle quadrant (see Figure 96ii) also add up to 34. Dürer's square therefore qualifies as a Gnomon Magic Square. The top left quadrant has the numbers $16 + 3 + 5 + 10 = 34$. The middle quadrant is $10 + 11 + 6 + 7 = 34$. The magic does not stop here. The two diagonals have even more to offer. The outmost pairs $(16, 1)$, $(13, 4)$ and the inner pairs $(10, 7)$, $(6, 11)$ all add up to 17, which is of course $\frac{34}{2}$.

Swimming Magic Squares

There are a total of 880 distinct 4×4 magic squares. If you add rotated and mirrored squares, then you receive a total of 7040. We can run a Python program to randomly select one of the 7040 that you can then swim, but since they are all permutations, we might as well use Dürer's original square. We can use the columns to determine the stroke and the rows to create variations (see Table 11). The strokes are defined as your sequence of preferred strokes, Nr1 being your favourite stroke. The numbers in the square tell you the number of laps to swim. This adds to $4 \times 34 = 136$ laps. In a 25 meters pool this adds up to 3400 meters.

You can download a Python program that randomly selects one of the 7040 programs and prints it to the command line here:

Table 11: Swimming Dürer's square.

	nr1	nr2	nr3	nr4
drill	16	3	2	13
pull	5	10	11	8
kick	9	6	7	12
swim	4	15	14	1

```
             ┌  16 D Nr 1 Any ◷0:00              1
             │   3 D Nr 2 Any ◷0:00              2
34 × 25 as  ┤   2 D Nr 3 Any ◷0:00              3
             └  13 D Nr 4 Any ◷0:00              4
             ┌   5 Nr 1 ◷0:00 Pullbuoy           5
             │  10 Nr 2 ◷0:00 Pullbuoy           6
34 × 25 as  ┤  11 Nr 3 ◷0:00 Pullbuoy           7
             └   8 Nr 4 ◷0:00 Pullbuoy           8
             ┌   9 K Nr 1 ◷0:00                  9
             │   6 K Nr 2 ◷0:00                 10
34 × 25 as  ┤   7 K Nr 3 ◷0:00                 11
             └  12 K Nr 4 ◷0:00                 12
             ┌   4 Nr 1 ◷0:00                   13
             │  15 Nr 2 ◷0:00                   14
34 × 25 as  ┤  14 Nr 3 ◷0:00                   15
             └   1 Nr 4 ◷0:00                   16
```

Figure 97: Swim Dürer's magic square swimming program. You can download this swimming program as a PDF from our repository.

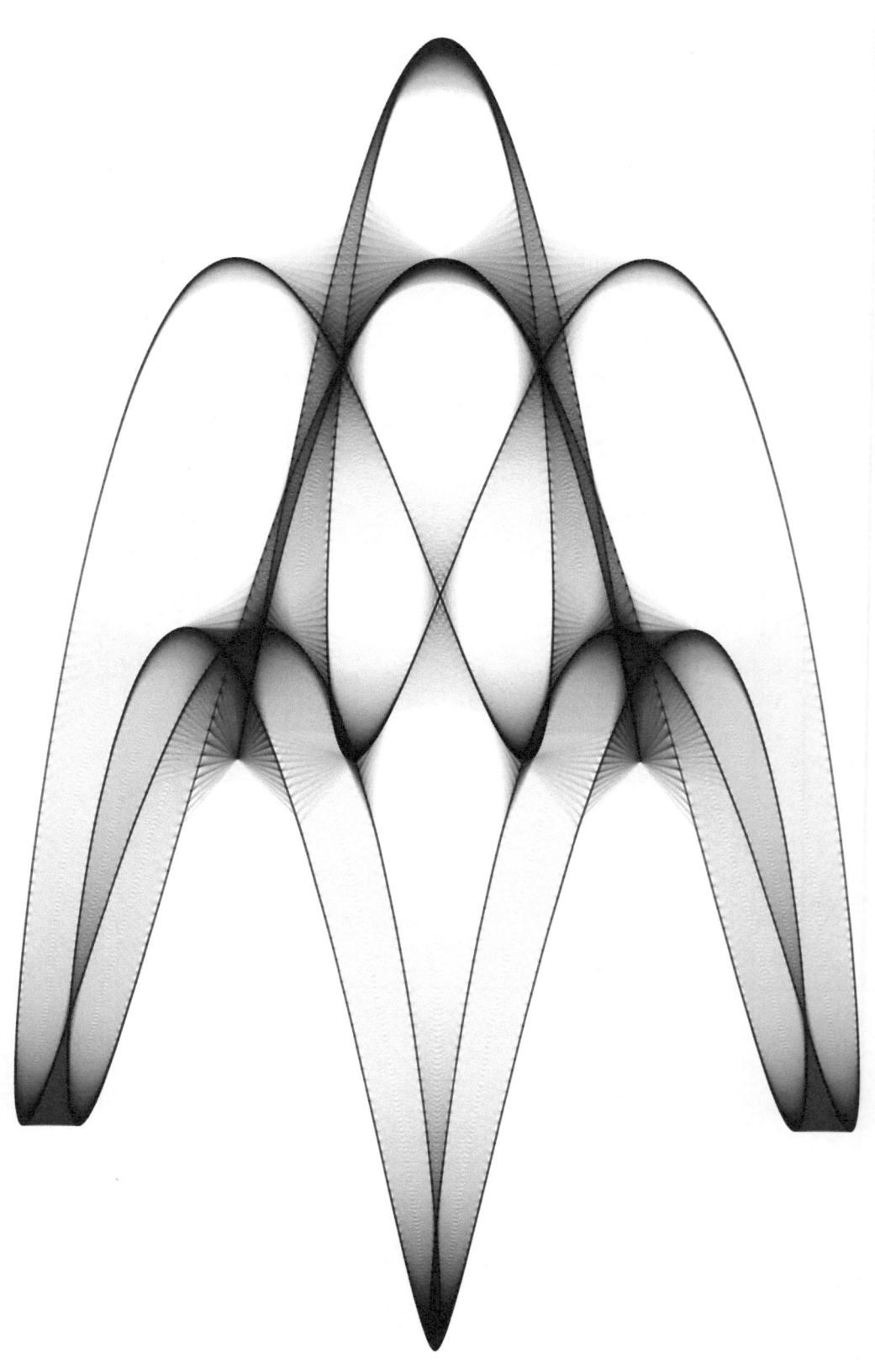

Multiplicative Persistence

Just like swimmers, numbers also have persistence. Multiplicative Persistence works like this. We take all the digits of a number and multiply them with each other.

$$327 \rightarrow 3 \times 2 \times 7 = 42$$
$$42 \rightarrow 4 \times 2 = 8 \tag{103}$$

The number 327 has a multiplicative persistence of two. It only took two steps to end in a single digit number (8) that can no longer be multiplied with another. Let's try a bigger number:

$$7615 \rightarrow 7 \times 6 \times 1 \times 5 = 210$$
$$210 \rightarrow 2 \times 1 \times 0 = 0 \tag{104}$$

Again, it only takes two steps to end in a single digit number (0). Feel free to try some numbers yourself, but you will notice that it is hard to find numbers with a long Multiplicative Persistence. One could start to become a bit more strategic in the search by not including certain digits, such as 5. If the numbers include a five, it will always take either one or two steps to produce a 0, which then ends the sequence. The current record holder for the smallest number with the highest persistence is:

$$277\,777\,788\,888\,899 \rightarrow 2 \times 7 \times 7 \times 7 \times 7 \times 7 \times 7 \times 8 \times 8 \times 8 \times 8 \times 8 \times 8 \times 9 \times 9 = 4\,996\,238\,671\,872$$
$$4\,996\,238\,671\,872 \rightarrow 4 \times 9 \times 9 \times 6 \times 2 \times 3 \times 8 \times 6 \times 7 \times 1 \times 8 \times 7 \times 2 = 438\,939\,648$$
$$438\,939\,648 \rightarrow 4 \times 3 \times 8 \times 9 \times 3 \times 9 \times 6 \times 4 \times 8 = 4\,478\,976$$
$$4\,478\,976 \rightarrow 4 \times 4 \times 7 \times 8 \times 9 \times 7 \times 6 = 338\,688$$
$$338\,688 \rightarrow 3 \times 3 \times 8 \times 6 \times 8 \times 8 = 27\,648$$
$$27\,648 \rightarrow 2 \times 7 \times 6 \times 4 \times 8 = 2\,688 \tag{105}$$
$$2\,688 \rightarrow 2 \times 6 \times 8 \times 8 = 768$$
$$768 \rightarrow 7 \times 6 \times 8 = 336$$
$$336 \rightarrow 3 \times 3 \times 6 = 54$$
$$54 \rightarrow 5 \times 4 = 20$$
$$20 \rightarrow 2 \times 0 = 0$$

This number has the currently highest known persistence of 11. There is currently no known number that has a persistence of 12 or higher. Table 12 shows the smallest number for each persistence level. There are some patterns that we can observe. First, the sequence of the digits does not matter, since they are being multiplied. Hence it makes most sense to put the digits in an increasing order. We also never observe more than one '2' or one '3' since two '2's could be replaced with one 4 and one 9 respectively. For large numbers, we end up with a few small digits at the beginning followed by series of 7, 8 and 9.

Calculating Persistence

We can easily calculate the Multiplicative Persistence of any number with a short computer program (see Listing 27). Try to enter a number yourself and see how persistent it is.

Table 12: This list showing the smallest number for each level of persistence is registered with the On-Line Encyclopedia of Integer Sequences as A022544 (`https://oeis.org/A003001`).

number	persistence
0	0
10	1
25	2
39	3
77	4
679	5
6 788	6
68 889	7
2 677 889	7
26 888 999	9
3 778 888 999	10
277 777 788 888 899	11

```python
def multiply_digits(n):
    result = 1
    while n > 0:
        result *= n % 10
        n //= 10
    return result


def multiplicative_persistence(n):
    persistence = 0
    while n >= 10:
        print(n)
        n = multiply_digits(n)
        persistence += 1
    return persistence


# Example usage
number = 99999
persistence = multiplicative_persistence(number)
print("The multiplicative persistence of " + str(number) + " is " +
 ↪  str(persistence) + " .")
```

Listing 27: Calculating the multiplicative persistence of any number.

Swimming Multiplicative Persistence

The goal of this program is to swim the currently most multiplicative persistent number: 277 777 788 888 899. We take the frequency of each digit as a repetition count and the digit itself for the number of laps. Hence we have 1×2 laps, 6×7 laps, 6×8 laps and 2×9 laps.

```
                    1 times 2 laps
─────────────────────────────────────────────────
1 × 50 FR                                          1
                    6 times 7 laps
         ─────────────────────────────────────────
         │ 100 FR Threshold                        2
6 × 175 as│ 50 FR Endurance                        3
         │ 25 FR Race Pace                         4
                    6 times 8 laps
─────────────────────────────────────────────────
6 × 200 IM ⏱0:15                                   5
                    2 times 9 laps
         ─────────────────────────────────────────
         │ 100 FR b3                               6
2 × 225 as│ 75 FR b5                               7
         │ 50 FR b7                                8
```

Figure 98: Multiplicative Persistence swimming program. You can download this swimming program as a PDF from our repository.

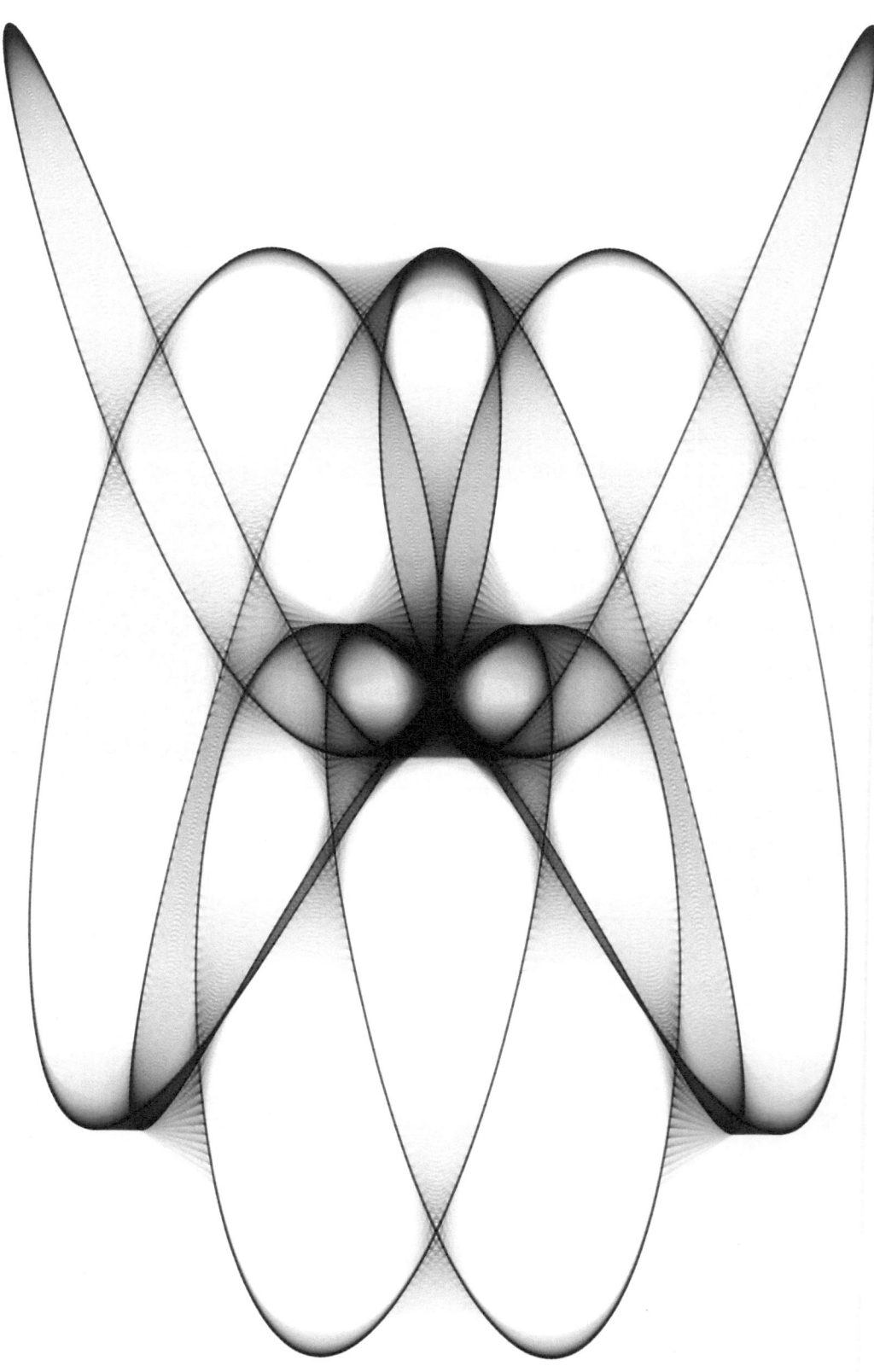

Square Sums

A swim training program is, at its heart, a sequence of laps with some rest in between. We then create patterns to give it a structure that swimmers can remember. Instead of a random sequence of laps, we swim 8 times 4 laps. There might be some training theory behind the structure but there is also a need for beautiful patterns. Let's consider one of the most simple sequences. Swim 1, 2, and 3 laps. Can you arrange this sequence so that every two adjacent ones sum to a square number? We can start with 1 and 3 since they sum up to the square number $1 + 3 = 4$ (see Equation 106). But the sum of 3 and 2 is 5, which is not a square number. The sequence breaks, and we are left with two disjointed segments.

$$\overbrace{1,3}^{4} \quad 2 \tag{106}$$

The smallest number that 2 could be paired with to get the square sum of 9 is 7 since $2 + 7 = 9$. Let's try a sequence of the numbers 1–7. We can arrange them into three disjointed segments:

$$\overbrace{1, \underbrace{3 ,6}_{9}}^{4} \quad \overbrace{2,7}^{9} \quad \overbrace{4,5}^{9} \tag{107}$$

A more generic solution to this problem is to create a graph that shows which pairs of numbers add up to square numbers (see Figure 100). The shortest sequence for which all numbers can be linked up is 1–15.

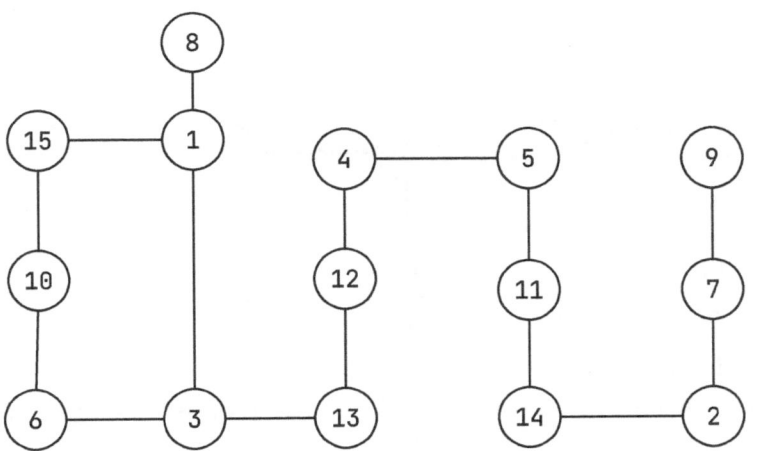

Notice that 1 can be paired with both 8 and 3 and thereby creates a loop (1, 3, 6, 10, 15). We need to find a path through this network so that each number is on its way only once. This is called a "Hamiltonian Path" (see Figure 101).

We can now arrange the numbers from 1–15 so that each adjacent number adds up to a square number:

Figure 99: This pattern and several other ideas in this book are based on the work of the mathematician and comedian Matt Parker (⋆1980). (source: Alasdhair Johnston)

Figure 100: Square sums network for the numbers 1–15.

Figure 101: Hamiltonian path for the numbers 1–15.

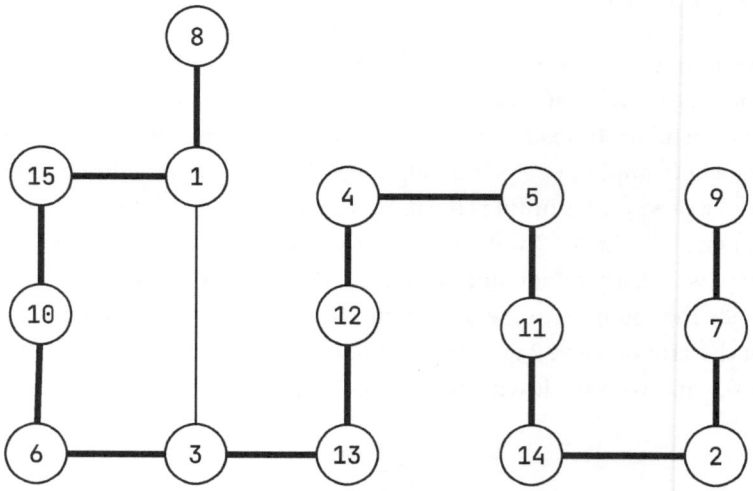

$$\overbrace{8, \underbrace{1}_{16}, \overbrace{15, 10, 6}^{9}, \underbrace{3}_{16}, \overbrace{13, 12, 4}^{25}, \underbrace{5}_{16}, \overbrace{11, 14, 2}^{9}, \underbrace{7}_{16}, 9} \tag{108}$$

While this is the shortest sequence, it is by far not the only sequence. We can simply extend this graph with the numbers 16 and 17 (see Figure 102).

Figure 102: Hamiltonian path for the numbers 1–17.

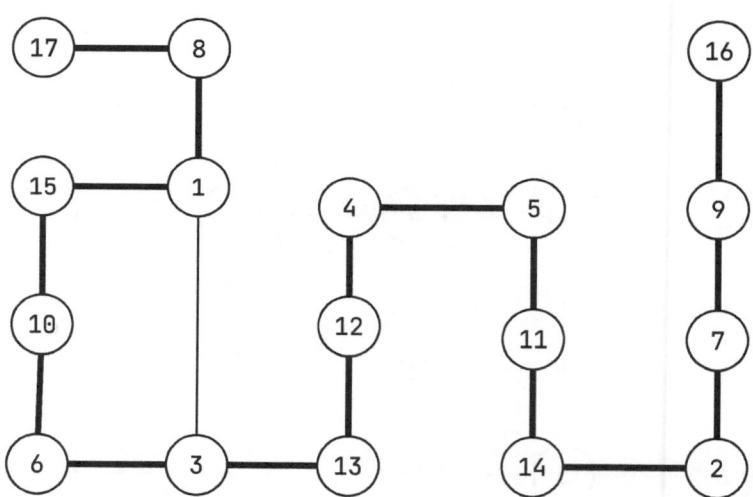

Once we add 18 to the sequence, we are stuck again since there is no Hamiltonian path for this network (see Figure 103). The path branches at 7, and we can never go through both 18 and [9,16]. If we add more numbers to the sequence, we can again draw a Hamiltonian Path. When we add all the numbers up to 23, it works, but the path breaks again for 24. All networks for the numbers 25 and above do have a Hamiltonian Path.

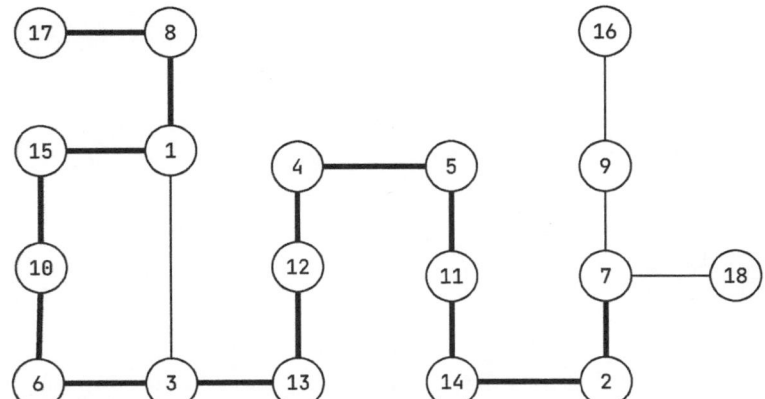

Swimming Square Sums

We can swim the smallest sequence of Square Sums by using the numbers as the count of laps and the squares as the stroke. Table 13 maps the four different squares we encounter with strokes.

Square	Stroke
4	Butterfly
9	Backstroke
16	Breaststroke
25	Freestyle

Table 13: Mapping squares to strokes.

We can then combine these strokes with the laps and get a swimming program (see Figure 104). Feel free to add different intensity levels, equipment or drills to make it even more interesting.

Figure 104: The Squared Sums swimming program. You can download this swimming program as a PDF from our repository.

8 laps BK ↻0:15			1
1 laps BR ↻0:15			2
15 laps FR ↻0:15			3
10 laps BR ↻0:15			4
6 laps BK ↻0:15			5
3 laps BR ↻0:15			6
13 laps FR ↻0:15			7
12 laps BR ↻0:15			8
4 laps BK ↻0:15			9
5 laps BR ↻0:15			10
11 laps FR ↻0:15			11
14 laps BR ↻0:15			12
2 laps BK ↻0:15			13
7 laps BR ↻0:15			14
9 laps BK ↻0:15			15

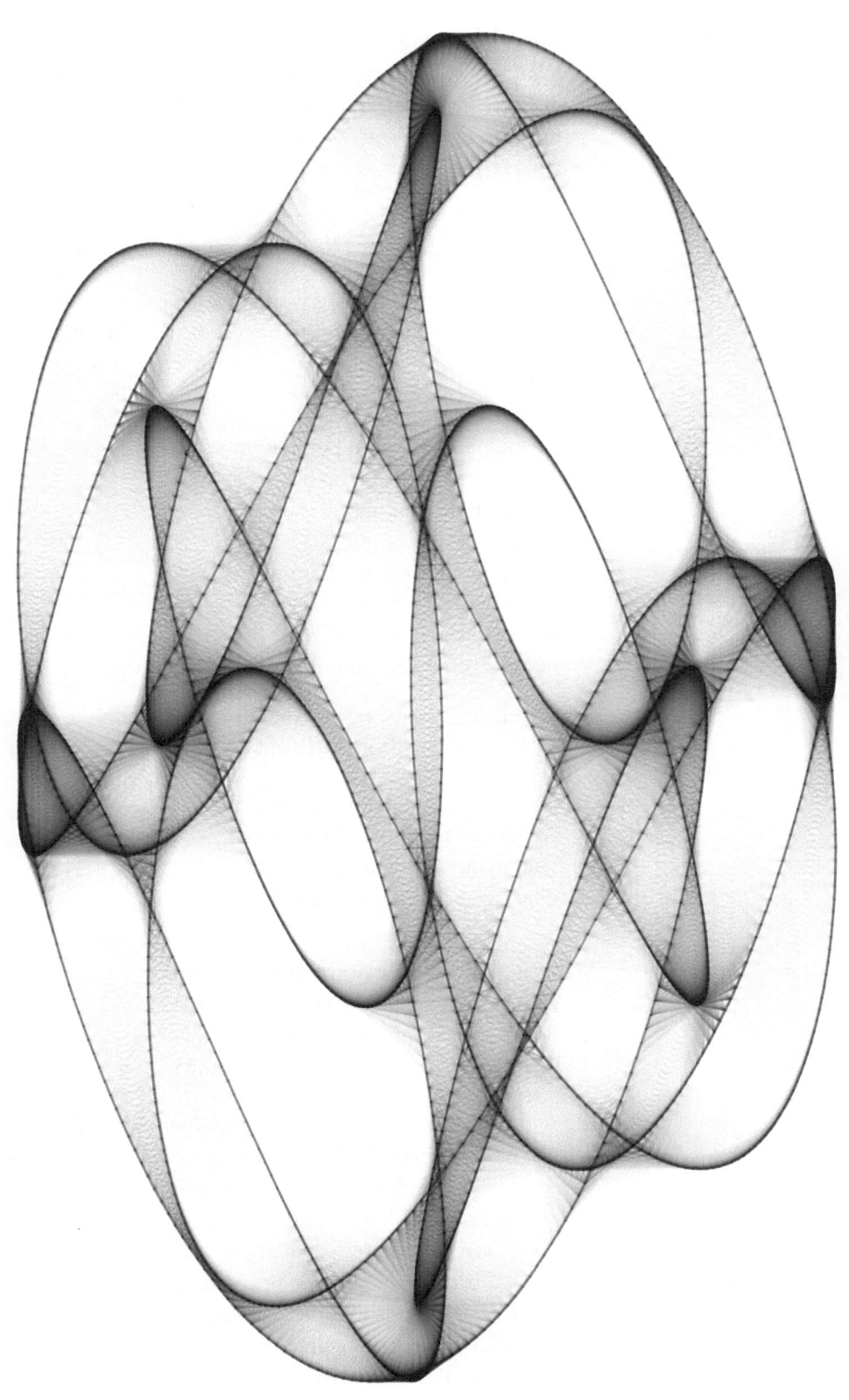

Repetition

Swimming is boring. Swimming up and down the lanes is different from riding down a mountain bike track, where all your focus has to be on the here and now. To keep themselves entertained, swimmers develop many methods to keep their minds busy. One of them is to recall a song or even hum a tune underwater. Some of us might be able to memorise the all the "American Pie" lyrics, while most of us can only sing along to the chorus:

So, bye, bye, Miss American Pie
Drove my Chevy to The Levee, but The Levee was dry...

An ideal song for swimming would, therefore, combine simple lyrics with a catchy tune. A true earworm. Let's start with the lyrics. We need to find the most repetitive song. You might remember such gems as "Pump Up The Jam" from the Belgian band Technotronic from 1989. The chorus is:

Yo, pump up the jam, pump it up
A-pump it up, yo, pump it
Pump up the jam, pump it up
A-pump it up, yo, pump it
Pump up the jam, pump it up
A-pump it up, yo, pump it
Pump up the jam, pump it
Pump it, pump it, pump it, yeah

This is already pretty repetitive, but is it more or less repetitive than the song "Funkytown" from the band Lipps, Inc. (1980)? Their chorus was:

Won't you take me to Funkytown?
Won't you take me to Funkytown?
Won't you take me to Funkytown?
Won't you take me to Funkytown?

Won't you take me down to Funkytown?
Won't you take me down to Funkytown?
Won't you take me down to Funkytown?
Won't you take me down to Funkytown?

To settle this debate, we need a measurement of repetitiveness. We don't have to look far, since the field of Information Theory has already explored this issue at length. One of the goals of Information Theory is to optimise the transmission of information by reducing the size of the transmission. If a message can be compressed, then it will take less information to transmit.

Let's go back to the times of telegrams and Morse code. Senders were charged by the length of their message, and hence code books were developed to replace common phrases with a letter sequence. An example is the Henry Roger's code book elegantly entitled, "The telegraph dictionary, and seamen's signal book, adapted to signals by flags or other semaphores; and arranged for secret correspondence, through Morse's electro-magnetic telegraph: for the use of commanders of vessels, merchants, &c" (Rogers,

1845). In it, the phrase "Captain is sick" is coded as "BTY" (see Figure 105). This cuts the message from 15 characters to only 3. This not only makes it cheaper to transmit but might also save you a few seconds before the captain decides to volunteer everybody on board to a swimming exercise.

B	T	T	1886	CAPTAIN.
				See Master or Commander.
B	T	U	1887	. . . I have seen the *captain.*
B	T	V	1888	. . . Inform the *captain.*
B	T	W	1889	. . . *Captain* is on board.
B	T	X	1890	. . . *Captain* is on shore.
B	T	Y	1891	. . . *Captain* is sick.
B	T	Z	1892	. . . *Captain* is well.
B	U	A	1893	. . . *Captain* is dead.

Figure 105: Code for "Captain is sick" from Henry Roger's code book (1845).

Modern communication still uses this technique. OMW, AFK or BRB have established meanings[26], but they are not part of a formal code book. Parents struggle to keep up with the communication of their teenagers.

[26] Ask any teenager if you are not familiar with these.

This code book approach therefore has three disadvantages. First, both sender and receiver need to have access to the same code book. There were and still are many of them around. During the height of the telegraph, there were not only Henry Roger's code book, but also Slater's Telegraphy Code (1916) and the Western Union Universal Codebook (1907). The sender and receiver first need to agree on which code book to use before starting to send the message. It would be possible to send the code book along with the message, but for short messages, this would increase the message size, not decrease it.

The second problem is that the code book only contains common phrases. Given the commercial context of many code books, the types of conversations possible were limited.

The third disadvantage is that the size of the code book can grow dramatically. Henry Roger's code book is already over 300 pages long. Looking up a coded message from such a book takes time and effort, reducing the overall efficiency of the communication.

A possible solution would have been to only use one code book that contains all the words in the dictionary. The most frequent words would be associated to shorter codes. While this more general approach works in principle, the number of characters saved per message is relatively low. What was necessary was a universal method to compress messages. This was achieved by Abraham Lempel and Jacob Ziv in 1978 (see Figure 106).

Lempel-Ziv algorithm

Their lossless compression algorithm LZ78 was published in 1978 by Lempel and Ziv (Ziv and Lempel, 1978). It was modified by Terry Welch in 1984. After Welch's publication, the algorithm was named LZW after the authors' surnames (Lempel, Ziv, Welch). It is in common use, such as for the compression in the Graphics Interchange Format (GIF) image file format. This format remains widely used for short animations.

Figure 106: Jacob Ziv (left, 1931-2023) and Abraham Lempel (right, 1936-2023), the inventors of the popular lossless data compression algorithm (source: Paul Urleib, Technion).

The general strategy of this compression algorithm is to create a code book specific for each message that is based on the repetitions in the message. The receiver would not know this code book, so it has to be sent together with the message itself. The challenge is to design an algorithm so that the combination of the encoded message with the code book is shorter than the original message itself. Otherwise it would not be a compression algorithm. The receiver can then use the transmitted code book to decode the message back to its original form.

Lempel and Ziv had an additional insight. They decided to encode the message on the fly. Meaning that the code book was incrementally created and transmitted in parallel to the submission of the encoded message. This enabled the receiver to already decode the message before the full combination of the encoded message and code book was received.

Let's consider a simple example. Matt wants to send the message AABABBABBAABA to Susanna. The algorithm starts at the beginning of the message and stops at the first letter it has not yet seen (Step 1 in Table 14).

A | ABABBABBAABA

In our case, this is the letter A. The | character indicates the stopping point in the sequence. It also has no entry for this letter in the code book and hence it submits this letter to Susanna. It then adds this letter to Matt's code book with A=1. Susanna receives the letter A and checks her code book. She has no entry for this and hence notes this letter in her code book as A=1. She then adds this letter to her received message.

Table 14: Steps in the LZW compression algorithm.

Step	Matt's code book		→	Susanna's code book		
	Sequence	Code		Sequence	Code	Message
1	A	1	A	A	1	A
2	AB	2	1B	AB	2	AAB
3	ABB	3	2B	ABB	3	AABABB
4	ABBA	4	3B	ABBA	4	AABABBABBA
5	ABA	5	2A	ABA	5	AABABBABBAABA

The algorithm moves forward to the next letter. It has already seen A, so it increases the sequence to the next letter B.

A | AB | ABBABBAABA

This is the first time the encoder has seen the sequence AB, so it stops. Instead of sending the sequence AB directly, it encodes it as what it knows is already in the code book of the receiver. Susanna's code book knows that A=1, so the encoder sends 1B. It then adds this sequence to its code book. Susanna receives the message 1B and knows that 1=A. The decoder replaces 1 with A and adds the sequence AB to her received message. So far her received message is AAB.

The algorithm then moves forward in the message. It has already seen the letter A so it moves forward to the next letter B. It has already seen the sequence AB so it moves forward one more letter. It has not yet seen the sequence ABB so it stops.

A | AB | ABB | ABBAABA

The algorithm knows that the sequence AB is already in Susanna's code as AB=2. It therefore sends the message 2B and adds the sequence ABB=3 to its code book (see step 3 in Table 14). Susanna receives the transmission

2B and knows that 2=AB so she replaces 2 with AB and adds the sequence ABB=3 to her code book. She also appends ABB to her received message which is AABABB at this point.

The algorithm then starts again. It has already seen the sequence A, AB and ABB. It moves forward one more letter to the sequence ABBA which it has not yet seen, so it stops.

A | AB | ABB | ABBA | ABA

As before, it transmits 3A and adds ABBA=4 to Matt's code book. Susanna knows that 3=ABB and appends ABBA to her received message. Next she adds ABBA=4 to her code book. The algorithm managed to compress a four letter sequence to a two letter transmission. It is becoming increasingly efficient.

In the last step, the algorithm starts with the sequence A and AB which it already knows. It moves forward one more letter until it encounters an unknown sequence ABA. It knows that Susanna's code book has the entry AB=3, so it sends 2A before adding ABA=5 to Matt's code book. Susanna receives 2A and replaces 2 with AB. It then appends ABA to the message and adds ABA=5 to her code book.

Admittedly, this message contained only two types of letters, A and B. A more conventional message is likely to contain at least the full set of ASCII characters (see page 11). Many messages are also much longer. Still, our example shows how the algorithm works in principle. It was able to reduce the 13 letters of the original message to a transmission of only 9 letters, a 30% compression rate. The longer and more repetitive the message is, the higher the compression rate will be. We can use this insight to determine the repetitiveness of music. We can use the compression rate as a measurement for the repetitiveness of a song.

When we apply the LZW compression algorithm to "Pump up the jam" we can achieve a size reduction of 85%. The crown for the most repetitive pop song belongs to Daft Punk's 1997 song "Around the World". It consists of 144 repetitions of the phrase "Around the World". It can be reduced by 98%.

While the lyrics are incredibly repetitive, the song was anything but boring. It became a major club hit globally and reached number one on the dance charts of many countries. Michel Gondry directed the iconic music video which features four robots, b-boys, skeletons, mummies and, most of all, swimmers (see Figure 107).

Figure 107: Synchronised swimmers in Daft Punk's video "Around the World". (Permission to use the image has been requested)

Figure 108: Prince Rogers Nelson (1958 – 2016) was a prolific artist and musician. He was also known as The Artist (Formerly Known as Prince) with his own name symbol. Unicode does not allow this symbol to be included in its character set, and hence, it will remain impossible to type this symbol on your computer. A workaround is based on the old-school emoji: `O(+>` (Permission to show the Love Symbol has been requested)

Figure 109: The Joy in Repetition swimming program. You can download this swimming program as a PDF from our repository.

It may fail to surprise you that the most repetitive pop song references swimming. The repetitive nature of lane swimming does give many an opportunity to let their mind wander. This does, however, conflict with the need to accurately count their progress. I personally never manage to accurately count 1500 meters freestyle. During competitions, I rely on the counters at the end of the lane and the whistle blown at the end. Still, as the artist formerly known as Prince pointed out in his 1990 song "Joy In Repetition":

> There's joy in repetition
> There's joy in repetition
> There's joy in repetition

Swimming Repetitions

The song "Around the World" repeats its only phrase 144 times. This is a good basis for the most repetitive swimming program. Luckily, $144 = 12^2$, which means that we can swim it as 12 times 12 times 50 meters, which results in a 7200 meters program. Alternatively, you can swim 12 times 12 times 25 meters to bring the total length down to 3600 meters (see Figure 109).

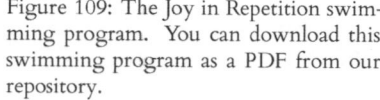

```
12 × 12 × 25 IM Order                                        1
```

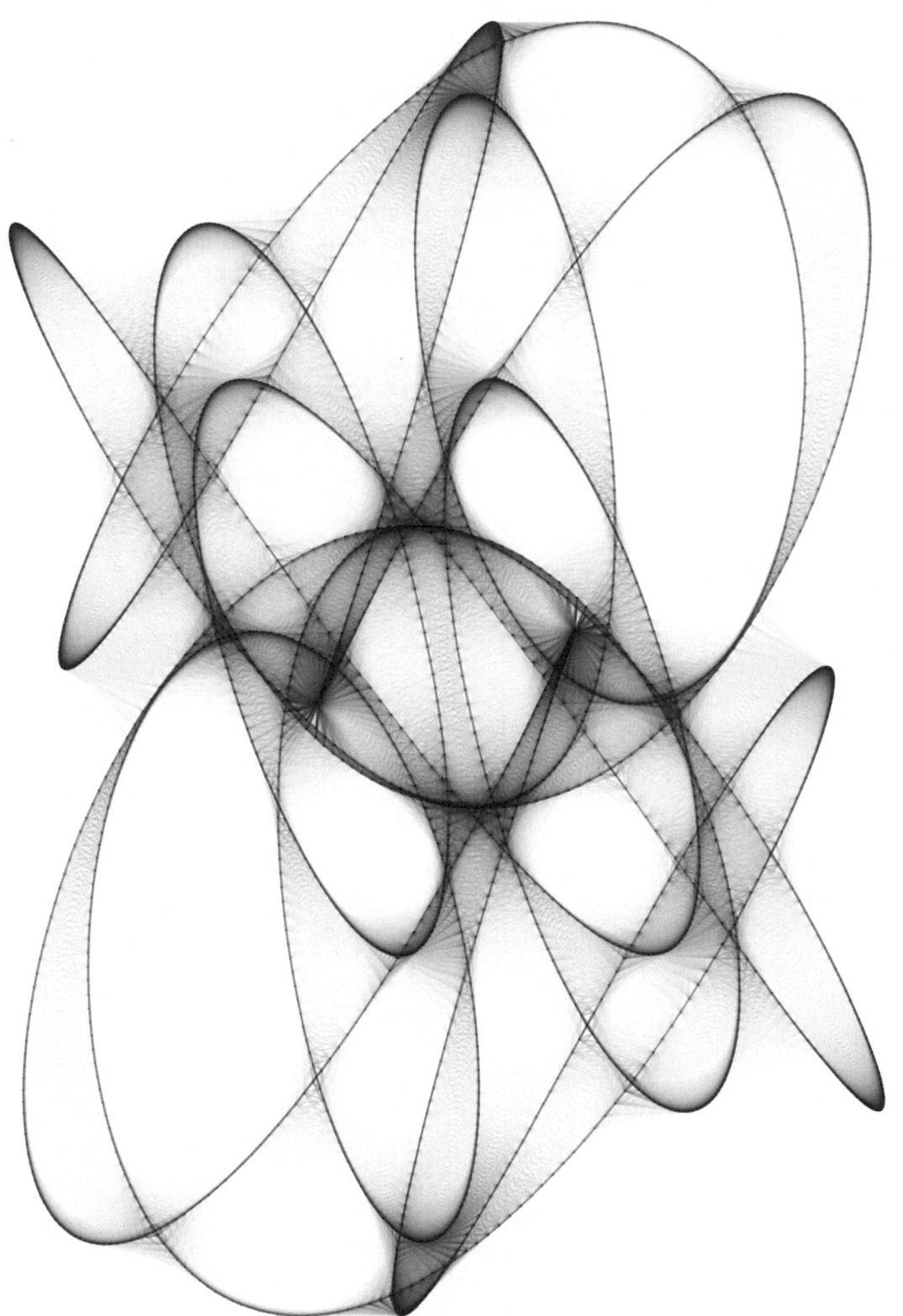

Hand Crossing

Swimming, like many other sports, is tightly linked to time. We measure success in how little of it we need to complete a distance. In training, we organise our exercise at intervals. A normal interval could be: swim four times 100 meters freestyle and start every 100 seconds. Many, but not all, pools have pace clocks that show only the seconds arm. This enables swimmers to time their intervals.

Almost all pools also have regular clocks with minutes and hours arms. Many swimmers might glance at this clock to determine either how long their suffering will continue or how long their bliss might last, depending on the program. While looking at such a clock, swimmers might wonder how often the minutes and hours arms meet. This will certainly happen at midnight and midday (see Figure 110).

Figure 110: Midnight and midday on a clock.

But what happens when the hour arm moves forward to 1? The minutes arm then also has to move forward to 5 minutes. But when the minutes arm moves forward by five minutes the hour arm will have moved as well (see Figure 111). So when exactly do they cross?

Figure 111: The minutes arm is chasing the hours arm.

The intuitive answer is twelve since there are twelve hours on a clock. But this would count both midday and midnight, although they are the same position of the arms. The right answer is eleven. 11 crossings in 12 hours. This is another example of the notorious Off-By-One or Picket Fence Error we encounter in more detail in Figure 122.

Time moves forward steadily. It does not speed up or slow down within our regular experiences on earth. This will be different at close to light speed, but we will neglect this for now. If there are 11 crossings in 12 hours, then the distance between the crossings must be evenly distributed. One hour has 3,600 seconds. This will result in:

$$\frac{12}{11} \times 3600 \tag{109}$$

To calculate this multiplication, we need to know what $\frac{12}{11}$ is. This superficially simple task has its own set of mysteries. Let's start at the beginning:

$$11 \overline{)12} \tag{110}$$

11 fits 1 time into 12 which leaves us with a remainder of 1:

$$\begin{array}{r} 1 \\ 11 \overline{)12} \\ \underline{11} \\ 1 \end{array} \tag{111}$$

11 does not fit into 1, so we have to draw down one zero:

$$\begin{array}{r} 1.0 \\ 11 \overline{)12.0} \\ \underline{11} \\ 1.0 \end{array} \tag{112}$$

11 does not fit into 10 either, so we have to draw down a second zero. 11 fits 9 times into 100.

$$\begin{array}{r} 1.\overline{09} \\ 11 \overline{)12.00} \\ \underline{11} \\ 1.00 \\ \underline{99} \\ 1 \end{array} \tag{113}$$

This leaves us again with a remainder of 1, which is the exact same situation we have encountered in Equation 111. This means that we entered an infinity loop. Instead of writing 1.090909090909..., we use the upper dash to express this repeating pattern: $1.\overline{09}$. We can never exactly know the decimal value since the pattern repeats indefinitely. But we are certain that it converges to the rational number of $\frac{12}{11}$. We can now multiply $1.\overline{09}$ by 3600 which results in:

$$\frac{12}{11} \times 3600 = 3927.\overline{27} \text{ seconds} \tag{114}$$
$$= 1 \text{ hour } 5 \text{ minutes } 27.\overline{27} \text{ seconds}$$

Achilles and the Tortoise

The minute arm chasing the hour arm is a mathematical paradox that was first described by Zeno of Elea (see Figure 112). In his series of paradoxes, Achilles is racing a tortoise. Since the tortoise is slower it gets a 100-meter head start. In the time it takes Achilles to reach the starting position of

Figure 112: Zeno of Elea (c. 490 BC – c. 430 BC).

the tortoise, it has already moved forward by 10 meters. Now Achilles has to spring another 10 meters to catch up. In that time, the tortoise has walked forward another meter. This keeps on happening over and over again. Achilles can never overtake the tortoise.

This does, of course, contradict our intuitive understanding of the world. Achilles will eventually be able to catch up to the tortoise. This apparent contradiction is Zeno's Paradox. During Zeno's lifetime, mathematics did not have the tools to solve the problem, and some argue that even modern mathematics hasn't.

Swimming the Crossings

If we divide 60 minutes into 11 segments, then we get a similar problem.

$$
\begin{array}{r}
5.\overline{45} \\
11\,\overline{)60.00} \\
55 \\
\hline
5.0 \\
4.4 \\
\hline
60 \\
55 \\
\hline
5
\end{array}
\tag{115}
$$

We can only approximate the solution to 5 minutes and 27 seconds as $0.\overline{45}$ minutes are around 27 seconds. We can then swim a training program that is divided into eleven periods.

Figure 113: The Hand Crossing swimming program. You can download this swimming program as a PDF from our repository.

5:27 Any ⟳0:15 Easy		1
5:27 FR ⟳0:15 70%		2
5:27 FR ⟳0:15 80%		3
5:27 FR ⟳0:15 90%		4
5:27 FR ⟳0:15 80%		5
5:27 FR ⟳0:15 70%		6
5:27 D FL Single Arm ⟳0:15 80%		7
5:27 BK ⟳0:15 80%		8
5:27 BR ⟳0:15 80%		9
5:27 FR ⟳0:15 80%		10
5:27 Any Easy		11

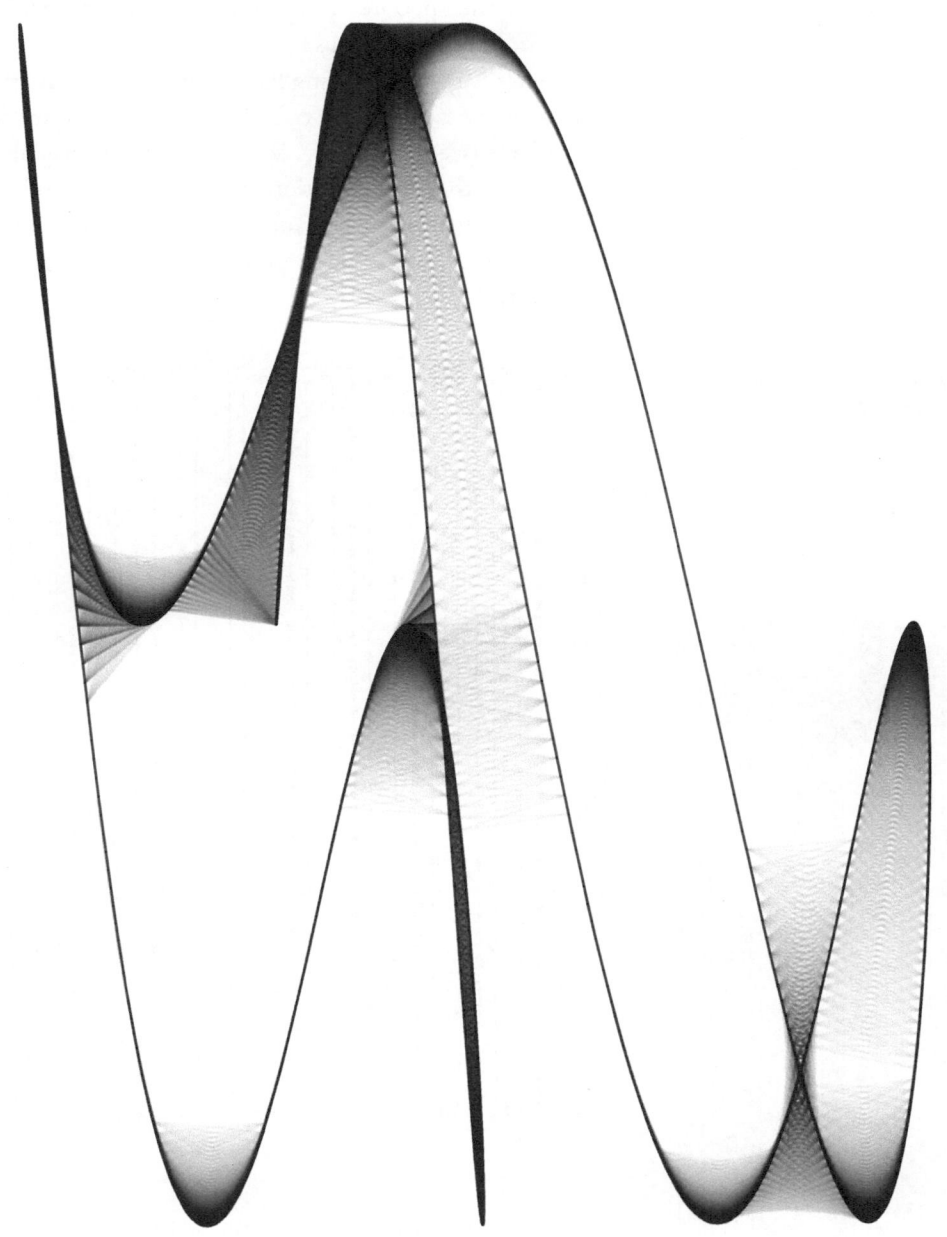

Round and Around

Usually, swimmers swim clockwise within a lane. This convention is particularly useful when swimming during public pool hours since this convention is widely accepted. It is, however, not the best way to organise lane swimming. When swimmers reach out of their lane, such as when swimming butterfly, their hands or legs might collide with swimmers coming in the opposite direction in the adjacent lane. Such collisions can be painful if not dangerous.

A better way of organising lane swimming is to alternate the directions. While the first lane swims clockwise, the second swims counter-clockwise, and so forth (see Figure 114). When swimmers meet along the lanes, they will swim in the same direction. This does not entirely prevent collisions, but the movement is in the same direction if they do occur. In some pools the direction in which to swim is even indicated on the starting blocks (see Figure 115).

Figure 114: All swimming in the clockwise directions (left) and alternating directions (right).

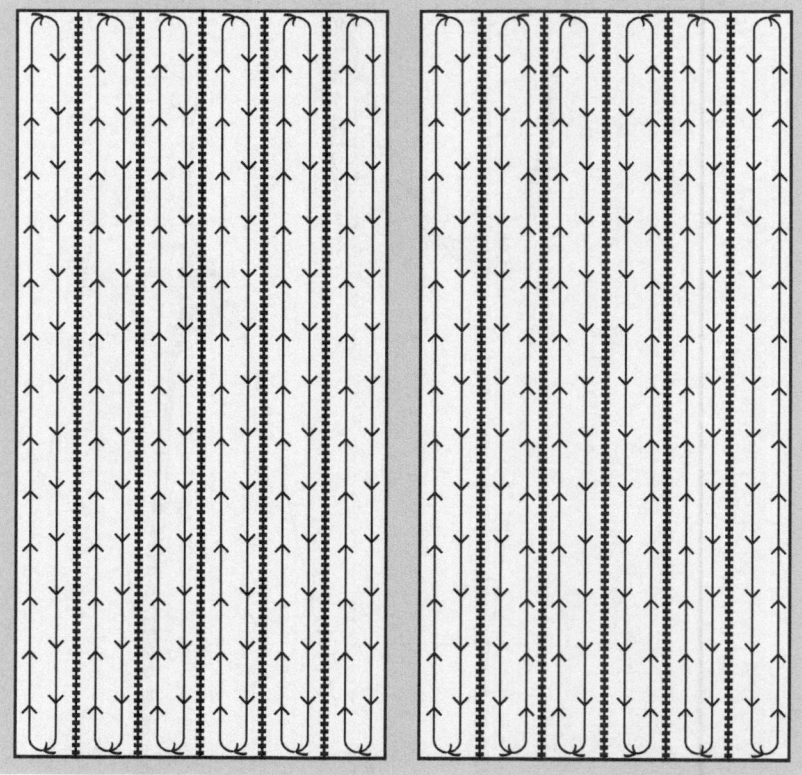

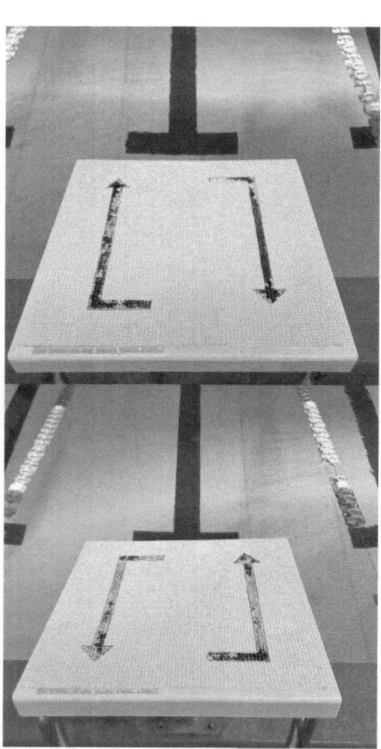

Figure 115: Swim directions posted on the start blocks.

We can abstract the paths that swimmers follow in the lanes into geometric circles (see Figure 116). When we have two identical circles that touch each other (Figure 116i), and both have their own rotational axis at their centres, then it follows that when one circle rotates once (without slipping), the other does so as well. Furthermore, if one disk is rolled along a line the length of its circumference, then it will rotate exactly once.

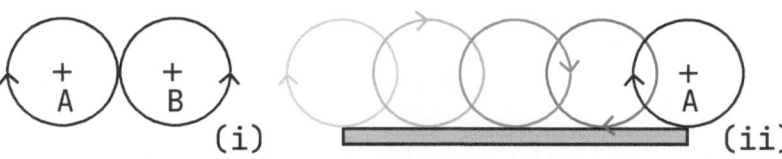

Figure 116: Circles rotating a single time.

But what happens when one circle is rolled around another (see Figure 117i)? Or what happens when the rolling circle B has a third of the radius of the stationary circle A (see Figure 117ii)? How often does the rolling circle B rotate around itself then?

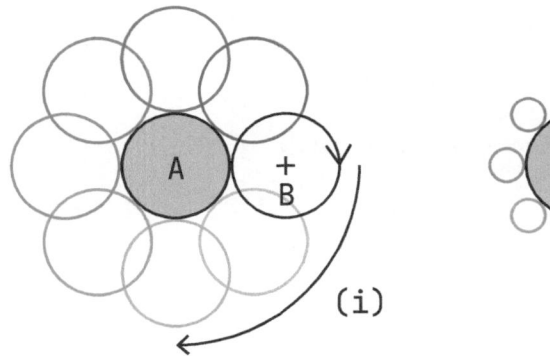

Figure 117: Two circles with identical radius (left) and two circles with the outer circle radius being $\frac{1}{3}$ of the inner (right).

The circumferences of circles A and B are:

$$C_A = 2\pi r_A$$
$$C_B = 2\pi r_B$$

(116)

If the two circles have the same radii, then they also have the same circumference. This might lead you to believe that circle B requires one rotation to roll around circle A.

You can easily try it out yourself. Simply take two identical coins and roll one around the other (see Figure 118). You will notice that it is two full rotations. This may lead us to believe that the number of rotations is twice that of the ratio of the radii.

Figure 118: Try rotating one coin around a second.

We would express this as:

$$R = 2 \times \frac{C_A}{C_B} \tag{117}$$

If you try out the same method with a circle B having a radius of $\frac{1}{3}$ that of circle A, you get a different result: four! To understand why, we first have to calculate the distance the centre of circle B travels (see Figure 119i).

Figure 119: The distance the centre of circle B travels around circle A (left) and a straight line (right).

(i)   (ii)

In 1982 the problem involving circles at a ratio of 3:1 was part of the SAT test. Unfortunately, the right answer of 4 was not listed as an option. Three students submitted proof to the College Board of why the answers to this SAT question were wrong: Sivan Kartha, Bruce Taub, and Doug Jungreis. The College Board had to admit their mistake and nullified the questions for all 300 000 students who took the test that year.

In our example, the centre of circle B travels:

$$2 \times \pi \times r$$
$$2 \times \pi \times (r_A + r_B) \tag{118}$$
$$2 \times \pi \times (3 + 1) = 8\pi$$

Circle B has a circumference of:

$$C = 2 \times \pi \times r_B$$
$$= 2 \times \pi \times 1 = 2\pi \tag{119}$$

The distance the centre travels is equal to the amount the circle rotates. To travel 8π the circle has to rotate:

$$\frac{8\pi}{2\pi} = 4 \tag{120}$$

If a circle is rotating on a straight line (see Figure 119ii) then the centre travels the same distance as the circumference, which always results in one rotation. The general solution for circles follows as:

$$\begin{aligned} R &= \frac{2 \times \pi \times (r_A + r_B)}{2 \times \pi \times r_B} \\ &= \frac{2 \times \pi \times r_A + 2 \times \pi \times r_B}{2 \times \pi \times r_B} \\ &= \frac{2 \times \pi \times r_A}{2 \times \pi \times r_B} + \frac{2 \times \pi \times r_B}{2 \times \pi \times r_B} \\ &= \frac{2 \times \pi \times r_A}{2 \times \pi \times r_B} + \frac{2 \times \pi \times r_B}{2 \times \pi \times r_B} \\ &= \frac{2 \times \pi \times r_A}{2 \times \pi \times r_B} + 1 \\ &= \frac{C_A}{C_B} + 1 \end{aligned} \tag{121}$$

If the ratio of circle A to circle B is 5:1, then the number of rotations necessary is:

$$R = \frac{2 \times \pi \times 5}{2 \times \pi \times 1} + 1 = 6 \tag{122}$$

This is another example of how our intuition can mislead us to an off-by-one error we encounter in more detail in Figure 122 on page 159.

Swimming Around

For the Round and Around swimming program we first swim the ratio of circle A and B before swimming the number of rotations of circle B around circle A.

Warm up	
400 Any Easy	1
First set	
100 FR	2
100 IM @_1:50 Endurance	3
200 Nr 4	4
Second set	
200 FR	5
100 IM @_1:50 Endurance	6
300 Nr 3	7
Third set	
300 FR	8
100 IM @_1:50 Endurance	9
400 Nr 2	10
Fourth set	
400 FR	11
100 IM @_1:50 Endurance	12
500 Nr 1	13

Figure 120: Round and Around swimming program. You can download this swimming program as a PDF from our repository.

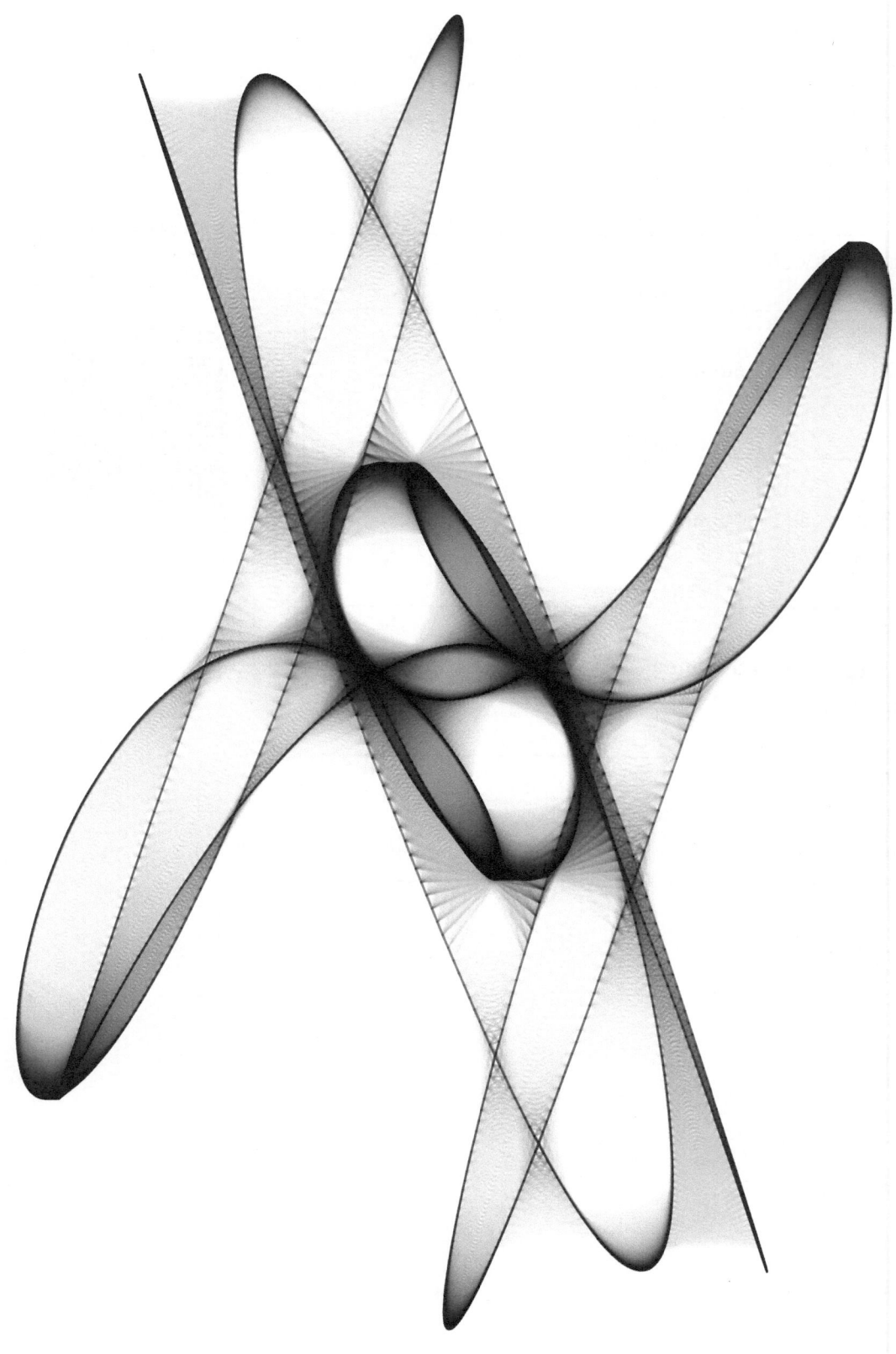

Time Pyramid

In 1993 the German city of Wemding celebrated its 1200th anniversary. They decided to start the long-term Zeitpyramide art project. Every decade they would place a concrete block on a base. After adding 1200 blocks, the pyramid would be complete (see Figure 121).

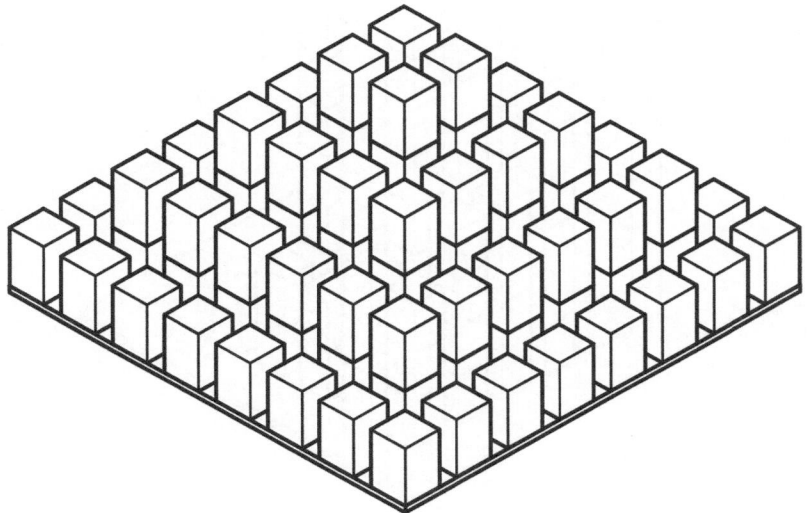

Figure 121: The original design of the Zeit-pyramide by Manfred Laber.

The only problem is that the pyramid will be completed in the year 3183. Which is only 1190 years in the future, not 1200. The project made a basic mathematical error, called the picket fence error. For a picket fence of n elements, we need to have $n + 1$ posts (see Figure 122). In swimming, we have a related relationship between laps and turns. To swim four laps you need three turns. In this case, you need $n - 1$ turns for n laps. More generally, this issue is referred to as the "Off By One Error".

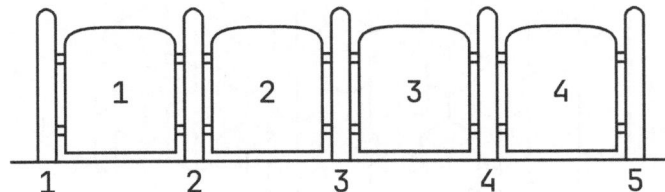

Figure 122: For a 4 element fence (n), we need 5 posts ($n + 1$).

Instead of waiting for the first decade to complete before placing the first block, they started immediately. This is like putting the first candle on the birthday cake on your child's actual day of birth.

Matt Parker (see Figure 99) pointed out this error during his visit to the placement of the fourth block in 2023. He also proposed an alternative design that would take 121 blocks to complete (see Figure 123). Unfortunately, his design is not a pyramid and would be 19.8 meters tall. That is certainly not safe in a storm.

Figure 123: Matt Parker's design for a 121 block Zeitpyramide.

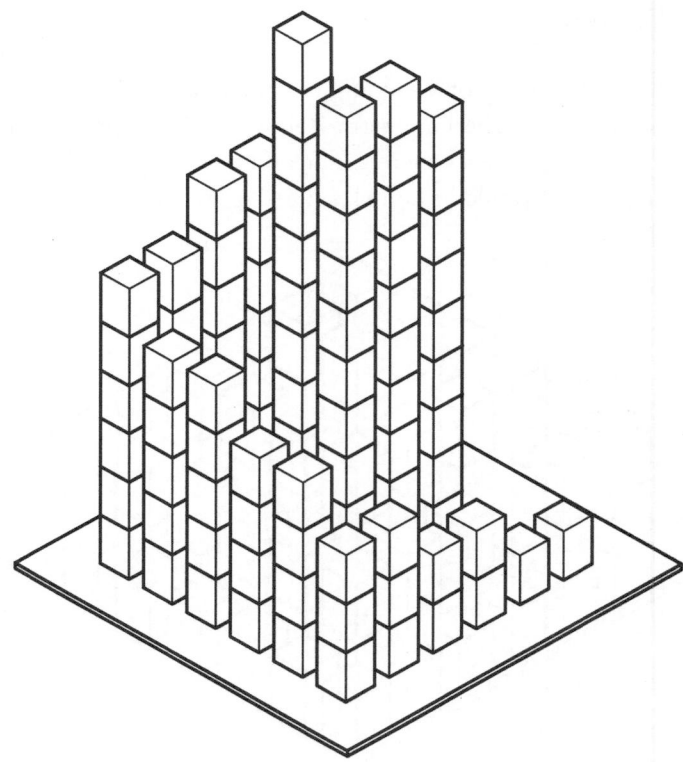

I took 121 of my beloved LEGO bricks and started on a seven by seven base. After some experimentation I came up with a beautiful pyramid that is only one block taller than the original design. It is still a proper pyramid with complete symmetry (see Figure 124).

Figure 124: My design of a 121 block Zeitpyramide.

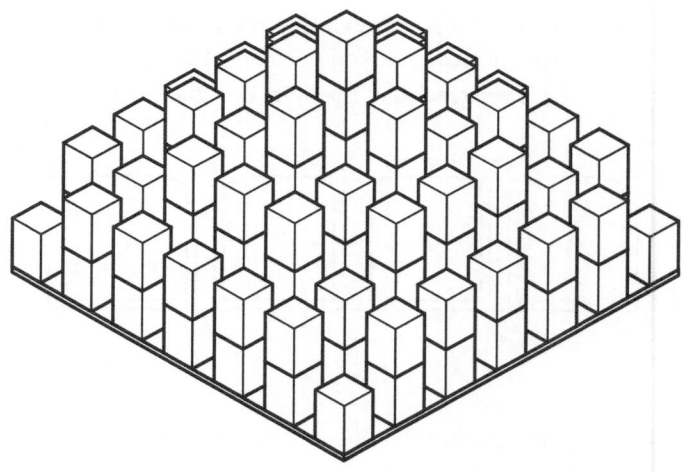

Figure 125: Manfred Laber (1932-2018) behind the Zeitpyramide model (copyright Stiftung Wemdinger Zeitpyramide).

We can only speculate what Manfred Laber, the artist, had in mind. According to Barbara Schlecht, head of the Zeitpyramide trust, Mr. Laber was fully aware of the consequences of his design. It is certainly much easier to design a sculpture with 120 bricks since it is divisible by 1, 2, 3, 4, 5, 6, 8, 10, 12, 15, 20, 24, 30, 40, 60 and 120. We could create a cuboid sculpture of $4 \times 6 \times 5$ blocks. 121, on the other hand, is only divisible by 1, 11 and 121.

We can speculate that Manfred Laber decided to sacrifice the opportunity to set a block at the beginning *and* at the end of the 1200 years period for having a direct relationship between the 120 blocks and the 1200 years. As shown above, there are alternative pyramid designs for 121 blocks. He could have also decided to only place the foundation in 1993.

Bonus

It is not uncommon for a construction project to steam ahead without all the details having been worked out. The Sydney Opera House is one of the most (in)famous examples. Its construction started in March 1959 without the design being completed. Many major questions were still unanswered. To justify the enormous expense of the build, the politicians, such as Jospeh Cahill, had to show quick progress. This resulted in much remedial work, once the structural integrity of the building was established.

In any case, this art project has become famous not for its original concepts, but for the controversy around its maths, which is unlikely to have been the intention of the artist.

Swimming the Zeitpyramide

Swimming the 121 block Zeitpyramide is best done row by row. Each block represents 25 meters (Figure 126).

Figure 126: The 121 block Zeitpyramide swimming program.

First row	
1 × **25** K Any	1
5 × **50** D Any Any	2
1 × **25** Any	3
Second row	
2 × **50** FR b3	4
3 × **75** FR b5	5
2 × **50** FR b3	6
Third row	
1 × **50** FR ⏱0:10 50%	7
2 × **75** FR ⏱0:15 75%	8
1 × **100** FR ⏱0:20 100%	9
2 × **75** FR ⏱0:15 75%	10
1 × **50** FR ⏱0:10 50%	11
Fourth row	
1 × **50** FL ⏱0:15	12
1 × **75** K Any ⏱0:15	13
1 × **100** BK ⏱0:15	14
1 × **125** K Any ⏱0:15	15
1 × **100** BR ⏱0:15	16
1 × **75** K Any ⏱0:15	17
1 × **50** FR ⏱0:15	18
Fifth row	
1 × **50** Not FR ⏱0:10 50%	19
2 × **75** Not FR ⏱0:15 75%	20
1 × **100** Not FR ⏱0:20 100%	21
2 × **75** Not FR ⏱0:15 75%	22
1 × **50** Not FR ⏱0:10 50%	23
Sixth row	
2 × **50** Nr 2	24
3 × **75** Nr 3	25
2 × **50** Nr 2	26
Seventh row	
1 × **25** K Any	27
5 × **50** D Any Any	28
1 × **25** Any	29

Handshake

When a group of swimmers come together for the first time to swim in a lane, they might decide to shake hands and introduce themselves. Shaking hands is also common amongst the recipients of medals after a competition. This polite gesture only takes a few moments, but it raises the question of how many hand shakes there will be. Let's consider the example of just two swimmers. Only one handshake is necessary. With three swimmers, three handshakes are in order, and with four swimmers already six handshakes occur (see Figure 127).

Figure 127: Handshakes necessary for 2-4 people.

$$H_2 = 1 \qquad H_3 = 3 \qquad H_4 = 6$$

When larger numbers of people meet, it does become impractical for everybody to shake hands. Still, it is interesting to work out the general solution to this problem. Before we start working on how many handshakes H are necessary for n people, let's get back to the specific problem of four people. We will consider the situation from the perspective of each of the four people.

- The first person shakes hands with everybody except for herself. She has to shake one less than the total number of people.

- The second person has already shaken hands with the first person but still has to shake everybody else's. Her number of handshakes is therefore two less than the total number of people.

- For n of people, this sequence continues until the second to last who only has to shake hands with the last person. In this case of four persons, the third person has already shaken hands with the first and second person. Her number of handshakes is three less than the total number of people.

For n people we can generalise that it will require:

$$H = (n-1) + (n-2) + (n-3) + \ldots + 2 + 1 \qquad (123)$$

This formula is a good start, but it would be better if we could simplify it even further. The trick here is to reverse this formula and add it to itself:

$$
\begin{array}{cccccc}
(n-1) & +(n-2) & +(n-3) & +\ldots & +2 & +1 \\
1+2 & & +\ldots & & +(n-3) & +(n-2) & +(n-1) \\
\hline
n+n & & +n & & +\ldots & +n & +n
\end{array}
$$

$$(124)$$

We get a series of n, and since our list ranges from $1 \ldots (n-1)$ it is clear that we have exactly $n-1$ times n. We can conclude that:

$$2H = n \times (n-1)$$
$$H = \frac{n \times (n-1)}{2} \qquad (125)$$

Figure 128: Johnny Weismuller (1904-1984) and Duke Kahanamoku (1890-1968) at the 1924 Olympic Games. After retiring from a successful swimming career, Johnny Weismuller abandoned handshakes in his new acting profession. He played Tarzan in the TV series and movies that involved a lot of swimming and wrestling crocodiles. He would use his signature yell instead of handshakes. Below you can see the spectral frequency of the yell which might help you to remember this. Interestingly, the yell is a palindrome. Meaning that it sounds the same, no matter if you play it forward or backward.

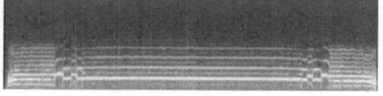

For a group of ten people this results in:

$$H = \frac{10 \times (10-1)}{2} = \frac{90}{2} = 45 \qquad (126)$$

With this general equation we can write out the sequence of handshakes:

$$1, 3, 6, 10, 15, 21, 28, 36, 45...$$

Triangular Numbers

This sequence has a direct relationship to triangular numbers. Triangular numbers count objects arranged in an equilateral triangle (see Figure 129).

The Triangular Numbers sequence is registered with the On-Line Encyclopedia of Integer Sequences as A000217 (https://oeis.org/A000217)

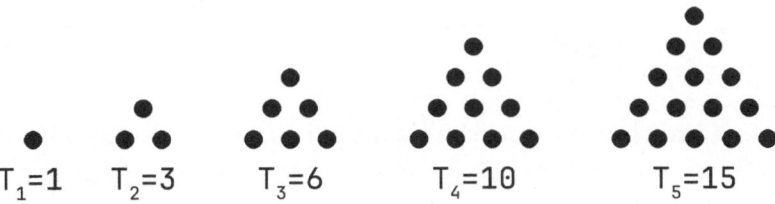

$T_1{=}1 \qquad T_2{=}3 \qquad T_3{=}6 \qquad T_4{=}10 \qquad T_5{=}15$

Figure 129: The first five triangular numbers.

The handshakes necessary for n people is equivalent to the $n-1$th triangular number. These can be described as:

$$T_n = \sum_{k=1}^{n} k = 1 + 2 + 3 + ... + n \qquad (127)$$

For multiplying a series of numbers, we do have the factorial expression denoted with the exclamation mark. We can consider the example of $4! = 1 \times 2 \times 3 \times 4 = 24$. Adding a sequence of numbers as seen in Equation 127 is slightly less established. Donald Knuth (see Figure 130) gave it the name "Termial Function" and associated the question mark symbol to it [27]. Hence, $3? = 1 + 2 + 3 = 6$.

We can write a short Python program to find n terms of the triangular numbers sequence (see Listing 28). Line 6 defines a simple loop that calculates each number of the sequence using the formula in Equation 125.

[27] Donald E Knuth. *The Art of Computer Programming: Fundamental Algorithms, Volume 1*. Addison-Wesley Professional, 1997. ISBN 9780201896831. URL https://search.worldcat.org/en/title/48246579

You can download this Python program from our repository.

```python
def triangular_sequence(n):
    # Initialize an empty array to store the sequence
    sequence = []
    # Generate the sequence and store it in the array
    for i in range(1, n + 1):
        triangular_number = (i * (i + 1)) // 2
        sequence.append(triangular_number)
    return sequence
# define number of terms
n = 10
# calculate sequence
triangular_numbers = triangular_sequence(n)
print("Sequence of", n, "triangular numbers:", triangular_numbers)
```

Listing 28: Searching for triangular numbers.

Figure 130: Donald Knuth (*1938)
(source: Jacob Appelbaum)

You might wonder how this detour to Triangular Numbers and the Terminal Function is relevant to this book. Donald Knuth also invented the the TeX typesetting system. The book you hold in your hands uses its offspring LaTeX (pronounced /ˈlaːtɛk/) to turn the text of the manuscript into a printable document.

> **Bonus**
>
> The LaTeX Project
>
> LaTeX is available for free for all computer platforms at `https://www.latex-project.org`. LaTeX uses another markup language (see Markup Languages on page 13). It combines text with instructions on how to layout the text. Here is an example:
>
> `This is the \textbf{best} equation: \( a^2+b^2=c^2 \)`
>
> Is transformed into:
>
> $$\text{This is the } \mathbf{best} \text{ equation: } a^2 + b^2 = c^2$$
>
> The tag `\textbf` stands for Text Bold Font and it marks text that should be shown in the bold typeface. The shorthand `\( \)` defines the start and end of a mathematical formula. The LaTeX bird logo was designed by Jonas Jacek.

Swimming Handshakes

We can use this simple program in Listing 28 to write a short Python program to generate a swimming program of any length using the triangular numbers sequence. The triangular number serves as the total length. To increase the variations in the program, we can subdivide each triangular number t_n into n (line 19) and $t_n - n$ (line 24) of Listing 29.

You can download this Python program from our repository.

```python
1   import swiML as swiML
2
3   def triangular_sequence(n):
4       # Initialize an empty array to store the sequence
5       sequence = []
6
7       # Generate the sequence and store it in the array
8       for i in range(1, n + 1):
9           triangular_number = (i * (i + 1)) // 2
10          sequence.append(triangular_number)
11
12      return sequence
```

```
13
14   def create_swiML_instructions():
15       my_instruction_list=[]
16       my_continue_list=[]
17
18       for i in range(0, nr_terms):
19           my_instruction_list.append(swiML.Instruction(
20                   length=('lengthAsLaps',i),
21                   stroke=('standardStroke','notFreestyle'),
22           ))
23           my_instruction_list.append(swiML.Instruction(
24                   length=('lengthAsLaps',triangular_numbers[i]-i),
25                   stroke=('standardStroke','freestyle'),
26           ))
27           # add instruction to the <continue> element.
28           my_continue_list.append(swiML.Continue(
29               instructions=my_instruction_list
30           ))
31           i+=1
32           my_instruction_list=[]
33       return my_continue_list
34
35   #writing the swiML program to disk
36   def write_program():
37       # warm up instructions
38       warmUp=swiML.Instruction(
39           length=('lengthAsDistance',200),
40           stroke=('standardStroke','any'),
41           intensity=('startIntensity',('zone','easy')),
42       )
43       # warm down instruction
44       warmDown=swiML.Instruction(
45           length=('lengthAsDistance',200),
46           stroke=('standardStroke','any'),
47           intensity=('startIntensity',('zone','easy')),
48       )
49       # assembly of the main instructions
50       myInstructions=[swiML.SegmentName('Warm
         ↪ Up'),warmUp,swiML.SegmentName('Triangular set')]
51       myInstructions.extend(create_swiML_instructions())
52       myInstructions.extend([swiML.SegmentName('Warm down'),warmDown])
53
54       # assemble the description of the swimming program
55       description_text="Swim the first "+str(nr_terms)+" terms of the
         ↪ triangular numbers sequence."
56
57       # create the program
58       simpleProgram=swiML.Program(
59           title='Triangular Numbers',
60           author=[('firstName','Christoph'),('lastName','Bartneck')],
61           programDescription=description_text,
62           poolLength='25',
63           creationDate='2024-08-20',
64           lengthUnit='meters',
65           hideIntro=False,
66           swiMLVersion='latest',
```

```
67        instructions=myInstructions
68      )
69      # write swiML XML to file
70      swiML.writeXML('handshake-program.xml',simpleProgram)
71
72    # define the number of terms
73    nr_terms=8
74    triangular_numbers = triangular_sequence(nr_terms)
75    # write the swiML program
76    write_program()
```

Listing 29: Triangular Numbers program.

Figure 131: The Triangular Numbers swimming program. You can download this swimming program as a PDF from our repository.

Warm Up			
200 Any Easy			1
Triangular set			
1 as	**0** laps Not FR		2
	1 laps FR		3
3 as	**1** laps Not FR		4
	2 laps FR		5
6 as	**2** laps Not FR		6
	4 laps FR		7
10 as	**3** laps Not FR		8
	7 laps FR		9
15 as	**4** laps Not FR		10
	11 laps FR		11
21 as	**5** laps Not FR		12
	16 laps FR		13
28 as	**6** laps Not FR		14
	22 laps FR		15
36 as	**7** laps Not FR		16
	29 laps FR		17
Warm down			
200 Any Easy			18

Knight's Tour

The tiles in the pool can be interpreted as a chess board. Swimmers move around on them, normally in a straight line like a rook. A rook can easily visit every square on the chessboard in a sequence. The same cannot be said about a knight. Its movement is more complex. A knight on a8 could move to b6 or c7 (see Figure 132).

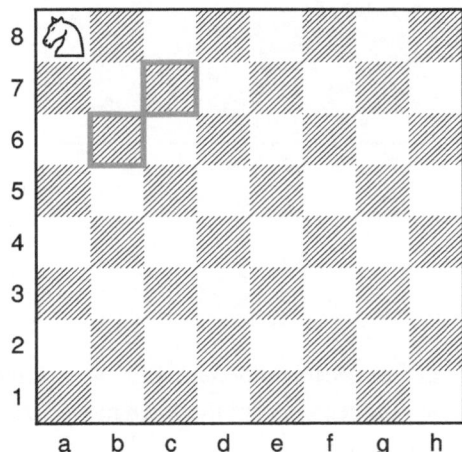

Figure 132: The possible moves for a knight on a8.

Let's move it to b6. From here, it could move to c8,d7,d5,c4 and a4 (see Figure 133).

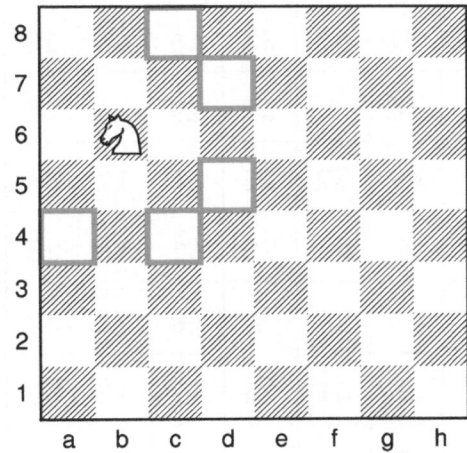

Figure 133: The possible moves for a knight on b6.

The goal is to find a sequence for the knight to move around the board so that it visits each field only once and returns in the end to its original position. This is called a closed knight's tour and there are several solutions. Figure 134 shows a solution that has another special attribute.

Figure 134: Magic knight's tour.

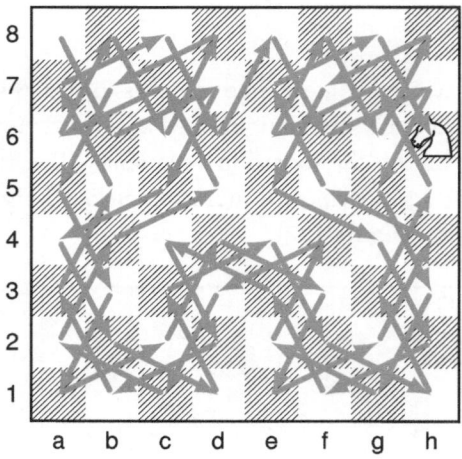

Magic Knight's Tour

This knight's tour has a magical component to it. We started in square a8, so this square gets the number 1. From there we moved to b6, which gets number 2 and so forth. When we put the number of the move into each square we get a board in which all squares have a number (see Figure 135).

Figure 135: This knight's tour results in a magic square.

1	48	31	50	33	16	63	18	260
30	51	46	3	62	19	14	35	260
47	2	49	32	15	34	17	64	260
52	29	4	45	20	61	36	13	260
5	44	25	56	9	40	21	60	260
28	53	8	41	24	57	12	37	260
43	6	55	26	39	10	59	22	260
54	27	42	7	58	23	38	11	260
260	260	260	260	260	260	260	260	

Adding up all rows and columns results in 260. Adding the quadrant also adds up to 260. The diagonals, however, do not add up to 260 as we have already seen for the magical squares in Figure 95 on page 131. We therefore only have a semi magic square. There is no fully magic square for an 8 × 8 board.

Bonus

Arguably the most minimalist chess set was designed by Bauhaus sculptor Josef Hartwig (1880–1956). In 1924 he created the Bauhaus Chess Model XVI. Each piece captures the essence of its movement in its most reduced form. It was originally manufactured from wood and sold in a cardboard box. Modern replicas are widely available.

Swimming the Knights Tour

Swimming this magic knight square can be accomplished by assigning a length and stroke to the rows and the style to the columns (see Figure 136). We get a comprehensive square that contains the dense information for this program.

	kick	drill	swim	pads	pull	swim	drill	kick	
25	1	48	31	50	33	16	63	18	butterfly
50	30	51	46	3	62	19	14	35	backstroke
75	47	2	49	32	15	34	17	64	breaststroke
100	52	29	4	45	20	61	36	13	freestyle
100	5	44	25	56	9	40	21	60	freestyle
75	28	53	8	41	24	57	12	37	breaststroke
50	43	6	55	26	39	10	59	22	backstroke
25	54	27	42	7	58	23	38	11	butterfly

Figure 136: The Magic Knight's tour program. You can download the Python program that finds magic knight's tours from our repository.

It would be possible to write this program in swiML, but due to the complex pattern, it would become a list of 64 instructions. The program shown in Figure 136 has much denser information and is therefore more suitable for this program.

Loops

Loops should not be confused with recursion (see page 26). Both can repeat instructions, but a recursion references itself. A loop simply repeats.

One of the main training structures is repetitions. swiML has the dedicated `<repetition>` element for this purpose. Repetitions are also essential to programming languages, and they are normally referred to as loops. Python has two types of loops: for loops and while loops. For now, we will focus on the for loops. Instead of using swiML's built-in `<repetition>` element which can be created in Python using `swiML.Repetition`, we can also directly use a for loop (see line 14 in Listing 30).

Find this Python program in our repository. You can download the Python program from our repository.

```python
1   import swiML as swiML
2
3   def simple_loop():
4       instruction_list=[]
5       for i in range (nr_loops):
6           instruction_list.append(swiML.Instruction(
7                   length=('lengthAsDistance',25),
8                   stroke=('standardStroke','butterfly'),
9           ))
10      return instruction_list
11
12  def pyramid_loop():
13      instruction_list=[]
14      for i in range(nr_loops):
15          instruction_list.append(swiML.Instruction(
16                  length=('lengthAsDistance',(i+1)*25),
17                  stroke=('standardStroke','notFreestyle'),
18          ))
19          instruction_list.append(swiML.Instruction(
20                  length=('lengthAsDistance',(nr_loops-i)*25),
21                  stroke=('standardStroke','freestyle'),
22          ))
23      return instruction_list
24
25  def nested_loop():
26      instruction_list=[]
27      for i in range(nr_inner_loops):
28          instruction_list.append(swiML.Instruction(
29                  length=('lengthAsDistance',(i+1)*25),
30                  stroke=('standardStroke','freestyle'),
31          ))
32          for j in range(i+1):
33              instruction_list.append(swiML.Instruction(
34                      length=('lengthAsDistance',100),
35                      stroke=('standardStroke','backstroke'),
36              ))
37      return instruction_list
38
39  #writing the swiML program to disk
40  def write_program():
41
42      # assembly of the main instructions
43      myInstructions=[swiML.SegmentName('Simple Loop')]
44      myInstructions.extend(simple_loop())
45      myInstructions.append(swiML.SegmentName('Double Loop'))
```

```
46    myInstructions.extend(pyramid_loop())
47    myInstructions.append(swiML.SegmentName('Nested Loop'))
48    myInstructions.extend(nested_loop())
49
50    # create the program
51    simpleProgram=swiML.Program(
52        title='Loops',
53        author=[('firstName','Christoph'),('lastName','Bartneck')],
54        programDescription='A program that showcases the loop
      ↪  function.',
55        poolLength='25',
56        creationDate='2024-08-10',
57        lengthUnit='meters',
58        hideIntro=True,
59        swiMLVersion='main',
60        instructions=myInstructions
61    )
62    # write swiML XML to file
63    swiML.writeXML('patterns/loops/loops-program.xml',simpleProgram)
64
65 # define the of number iterations for the loops
66 nr_loops=4
67 nr_inner_loops=3
68 # write the swiML program
69 write_program()
```

Listing 30: Loops code.

The simple loops program requires more lines in the rendered PDF and has therefore no immediate advantage (see lines 1–4 in Figure 138).

The true power of a full programming language, such as Python, comes to the forefront when considering using the counter in the loops to manipulate the swimming instructions. Loops require a counter variable to keep track of how many times the loop was executed. When we look at the pyramid loop (lines 12–23 in Listing 30) we see the counter variable i. In each iteration of the loop, two instructions are written. The first increases from 25–100 meters (not freestyle), which can be expressed as (i+1)*25. Notice that the loop starts counting at zero, and hence we have to start the calculation of the distance with i+1. The second instruction decreases from 100–25 meters (freestyle). This can be expressed as (nr_loops-i)*25.

The resulting training pattern (see lines 5–12 in Figure 138) cannot be modelled with swiML's <repetition> element. But this is not the end of what can be accomplished with loops. Similar to the <instruction> elements being able to be nested, so can loops. The nested_loop() function (see lines 25–37 in Listing 30) shows two nested loops. The first loop is again increasing the distance with (i+1)*25. A second loop is nested in the first. Its number of iterations j is based on the counter variable of its enclosing loop's counter variable i. This results in a training pattern (see lines 13–21 in Figure 138) in which the count of 100 meters backstroke increases with every iteration of the outer loop.

Figure 137: Augusta Ada King, Countess of Lovelace (1815-1852), daughter of Lord Byron, is considered by many as one of the first programmers. She worked on software that used nested loops for Charles Babbage's mechanical general-purpose computer. The machine was never completed and so her thoughts remained unapplied during her lifetime.

Figure 138: Simple, double and nested loops.

Simple Loop		
25 FL		1
25 FL		2
25 FL		3
25 FL		4
Double Loop		
25 Not FR		5
100 FR		6
50 Not FR		7
75 FR		8
75 Not FR		9
50 FR		10
100 Not FR		11
25 FR		12
Nested Loop		
25 FR		13
100 BK		14
50 FR		15
100 BK		16
100 BK		17
75 FR		18
100 BK		19
100 BK		20
100 BK		21

Swimming Loops

Using the power of loops in Python allows us to write programs concisely that create complex patterns when written out in swiML. Listing 31 shows nested loops (lines 3–25), which are responsible for the complex patterns shown in Figure 139. Again, the rendered program does not contain `<repetition>` elements. The repeating nature of the pattern is solely based on the nested loops in the Python program.

Find this Python program in our repository. You can download the Python program from our repository.

```python
import swiML as swiML


def nested_loop():
    instruction_list=[]
    for i in range(nr_inner_loops):
        instruction_list.append(swiML.Instruction(
                length=('lengthAsDistance',(i+1)*25),
                stroke=('standardStroke','breaststroke'),
```

```
 9                 rest=('afterStop',"PT0M15S"),
10                 breath=i+1
11             ))
12         for j in range(i+1):
13             instruction_list.append(swiML.Instruction(
14                 length=('lengthAsDistance',100),
15                 stroke=('standardStroke','freestyle'),
16                 rest=('sinceStart',"PT1M"+str((40+(j*5)))+"S"),
17                 intensity=('startIntensity',('percentageEffort',
18                     (100-(10*nr_inner_loops))+(j*10)))
19             ))
20             instruction_list.append(swiML.Instruction(
21                 length=('lengthAsDistance',(j+1)*25),
22                 stroke=('standardStroke','backstroke'),
23                 rest=('sinceStart',"PT0M"+str((j+1)*35)+"S")
24             ))
25     return instruction_list
26
27 #writing the swiML program to disk
28 def write_program():
29     # assembly of the main instructions
30     # warm up instructions
31     warmUp=swiML.Instruction(
32         length=('lengthAsDistance',400),
33         stroke=('standardStroke','any'),
34         intensity=('startIntensity',('zone','easy')),
35     )
36     # assembly of the main instructions
37     myInstructions=[swiML.SegmentName('Warm Up'),warmUp]
38     myInstructions.append(swiML.SegmentName('Nested Loop'))
39     myInstructions.extend(nested_loop())
40     myInstructions.extend([swiML.SegmentName('Warm down'),warmUp])
41     # create the program
42     simpleProgram=swiML.Program(
43         title='Nested Loops',
44         author=[('firstName','Christoph'),('lastName','Bartneck')],
45         programDescription='A program that showcases nested loops.',
46         poolLength='25',
47         creationDate='2024-08-10',
48         lengthUnit='meters',
49         hideIntro=False,
50         swiMLVersion='main',
51         instructions=myInstructions
52     )
53     # write swiML XML to file
54     swiML.writeXML('loops-program-swim.xml',simpleProgram)
55
56 # define the number of iterations for the loops
57 nr_loops=4
58 nr_inner_loops=4
59 # write the swiML program
60 write_program()
```

Listing 31: Python program that uses loops to generate a training program.

Figure 139: The Loops swimming program. You can download this swimming program as a PDF from our repository.

Warm Up	
400 Any Easy	1
Nested Loop	
25 BR ↻0:15 b1	2
100 FR @_1:40 60%	3
25 BK @_0:35	4
50 BR ↻0:15 b2	5
100 FR @_1:40 60%	6
25 BK @_0:35	7
100 FR @_1:45 70%	8
50 BK @_1:10	9
75 BR ↻0:15 b3	10
100 FR @_1:40 60%	11
25 BK @_0:35	12
100 FR @_1:45 70%	13
50 BK @_1:10	14
100 FR @_1:50 80%	15
75 BK @_1:45	16
100 BR ↻0:15 b4	17
100 FR @_1:40 60%	18
25 BK @_0:35	19
100 FR @_1:45 70%	20
50 BK @_1:10	21
100 FR @_1:50 80%	22
75 BK @_1:45	23
100 FR @_1:55 90%	24
100 BK @_2:20	25
Warm down	
400 Any Easy	26

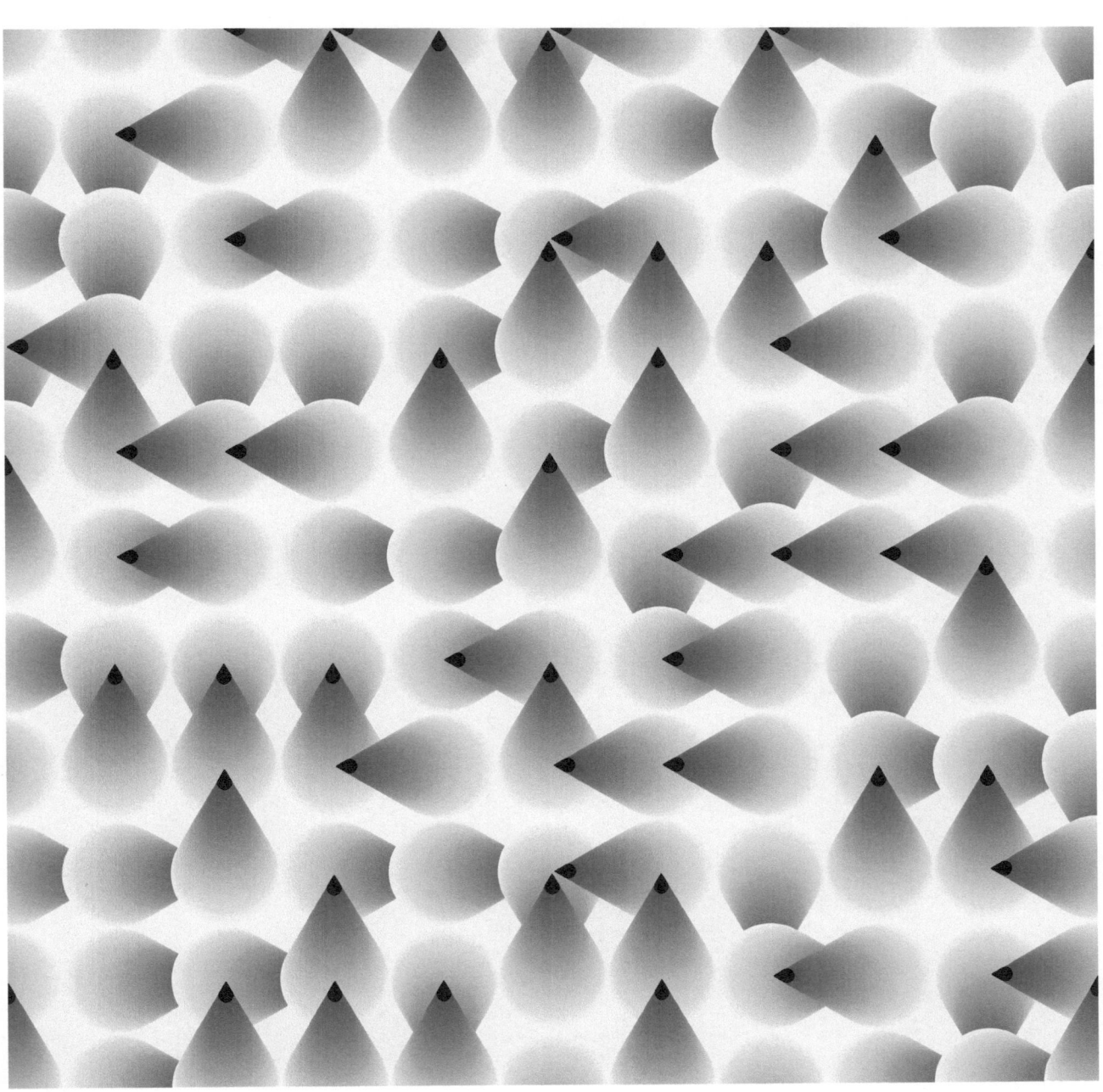

Regular Programs

Summary

This chapter showcases how swiML can be used to author interesting swim training programs. These traditional programs provide inspiration for when you want to focus on physical exercise to give your head a rest.

Besides the programs listed in this chapter, there are also over a hundred swimming programs available from the Jasi Masters swimming club[28]. The programs were authored by their coach, Matt Nash. They are normally targeted at one hour and are around 3 000 meters.

Figure 140: The Jasi Masters Swimming Club was established in 1989 in Christchurch, New Zealand.

[28] https://bartneck.github.io/swiML/jasiMasters/

Figure 141: Regular program 1.

Regular Program 1

Christoph Bartneck

A simple program mixing longer aerobic
freestyle with shorter sprints.

⤳Date:01 June 2024
⤳Pool Size:25
⤳Units:meters
⤳Length:3100 meters / 124 Laps

Warm up		
300 Any Easy		1
300 as	**100** K Any	2
	100 D Any Any	3
	100 Any	4
First set		
400 FR Pads Pullbuoy		5
4 × **25** IM Order @_0:30		6
2 × Easy…Race Pace **300** FR ☉0:45		7
2 × **50** Nr 1 @_1:00		8
3 × Easy…Race Pace **200** FR @_3:15		9
2 × **50** IM Overlap @_1:00		10
4 × Easy…Race Pace **100** FR @_1:35		11
4 × **25** Nr 1 @_0:30		12
Warm down		
100 Any Easy		13

made with: **swiML**

Regular Program 2

Christoph Bartneck

The main set of this program combines a decreasing freestyle instruction with shorter sprints.

Figure 142: Regular program 2.

⤳**Date:**04 June 2024
⤳**Pool Size:**25
⤳**Units:**meters
⤳**Length:**3100 meters / 124 Laps

Warm up	
400 Any Easy	1
4 × Fins ⌠ **50** K BK	2
100 as ⌜ **50** D IM Order Any	3
⌞ **50** FR	4

First set	
250 FR ↻0:30 Pads Pullbuoy	5
50 K Any ↻0:20 Race Pace	6
3 × **150** FR ↻0:20	7
50 BK ↻0:20 Race Pace Pullbuoy	8
50 FR ↻0:20	9
50 Nr 1 ↻1:00 Race Pace	10

Second set	
8 × **25** IM Order @_0:45	11

Warm down	
100 Any Easy	12

made with: **swiML**

Figure 143: Regular program 3.

Regular Program 3

Christoph Bartneck

The main set of this program combines a freestyle pyramid with shorter sprints in IM or Nr 1.

⇝Date:01 July 2024
⇝Pool Size:25
⇝Units:meters
⇝Length:3100 meters / 124 Laps

Warm up			
400 Any Easy			1
300 *Second round not freestyle* as	**50** K FR		2
	50 D FR Any		3
	50 FR		4
First set			
400 FR Pads Pullbuoy			5
4 × **25** IM Order @_0:30			6
2 × Threshold…Endurance **300** FR ↻0:45			7
2 × **50** Nr 1 @_1:00			8
3 × Easy…Endurance **200** FR ↻3:15			9
2 × **50** IM Overlap @_1:00			10
4 × Easy…Race Pace **100** FR @_1:35			11
4 × **25** Nr 1 @_0:30			12

made with: **swiML**

Regular Program 4

Christoph Bartneck

A freestyle set with a main block of 100 meters with increasing intensity.

⤳**Date:**04 June 2024
⤳**Pool Size:**25
⤳**Units:**meters
⤳**Length:**3000 meters / 120 Laps

Warm up	
4 × **100** FR	1
4 × **50** IM Overlap	2
First set	
4 × 60…90% 4 × **100** FR @_1:35 *Extra 1:00 rest*	3
6 × ⎡ **50** FR @_1:00 Max	4
⎣ **50** Not FR @_1:00 Easy	5
Warm down	
200 Any Easy	6

made with: **swiML**

Figure 145: Regular program 5.

Regular Program 5

Christoph Bartneck
This set is focused on the number five.

⤳Date:15 July 2024
⤳Pool Size:25
⤳Units:meters
⤳Length:3000 meters / 120 Laps

Warm up			
500 as	**100** FR		1
	25 Not FR		2
First set			
4 ×	**150** FR ⊙0:20 Endurance		3
3 ×	**150** BK ⊙0:20 Endurance		4
2 ×	**150** BR ⊙0:20 Endurance		5
1 ×	**150** D FL Single Arm		6
Second set			
5 ×	**100** FR ⊙0:05 Race Pace		7
	50 FR ⊙0:05 Race Pace		8
	50 FR ⊙0:05 Race Pace		9

made with: **swiML**

Regular Program 6

Figure 146: Regular program 6.

Christoph Bartneck

The sixth regular program celebrates the
perfect number six.

⤳Date:06 June 2024
⤳Pool Size:25
⤳Units:meters
⤳Length:3600 meters / 144 Laps

Warm up			
600 as	3 ×	**100** FR	1
		100 Nr 1	2
First set			
2 × **300** FR Pads Pullbuoy			3
3 × **200** FR Pads			4
6 × **100** FR Pullbuoy			5
12 × **50** FR ↻0:30 Race Pace			6
Warm down			
600 as	2 ×	**150** Nr 1 Easy	7
		150 BK Easy	8

made with: **swiML**

Figure 147: Regular program 7.

Regular Program 7

Christoph Bartneck

This program mixes freestyle with individual medley, easy with fast swimming.

⤳Date:15 July 2024
⤳Pool Size:25
⤳Units:meters
⤳Length:3300 meters / 132 Laps

Warm up		
	50 K Any	1
300 as	**50** D Any Any	2
	50 Any	3
First set		
	300 FR ⟳0:30 Threshold	4
3 ×	**200** IM ⟳0:20 Endurance	5
	100 K Any Easy	6
Second set		
	50 FR ⟳0:20 Easy	7
	100 IM ⟳0:10 Race Pace	8
3 ×	**50** Any ⟳0:20 Easy	9
	75 FR ⟳0:10 Race Pace	10
	50 Any ⟳0:20 Easy	11
Warm down		
225 Any Easy		12

made with: **swiML**

Regular Program 8

Christoph Bartneck
This program mixes freestyle with increasingly
fast individual medley. The second set focuses
on stroke counts.

⤳**Date:**18 July 2024
⤳**Pool Size:**25
⤳**Units:**meters
⤳**Length:**3400 meters / 136 Laps

	Warm up	
3 × **200** as	**1** Any ⟳0:20 Easy	1
	1 K Any ⟳0:20 Easy	2
	1 D Any Any ⟳0:20 Easy	3

	First set	
500 FR ⟳0:15		4
100 IM @_1:45 60%		5
400 FR ⟳0:15		6
100 IM @_1:45 70%		7
300 FR ⟳0:15		8
100 IM @_1:45 80%		9
200 FR ⟳0:15		10
100 IM @_1:45 90%		11
100 FR ⟳0:15		12
100 IM @_1:45 100%		13

	Second set	
100 FR Easy *Count strokes*		14
6 × **100** FR @_1:45 *Reduce stroke count*		15
100 Any Easy		16

made with: **swiML**

Figure 149: Regular program 9.

Regular Program 9

Christoph Bartneck
This program focuses on kicking.

⤳Date:19 July 2024
⤳Pool Size:25
⤳Units:meters
⤳Length:3200 meters / 128 Laps

Warm up	
300 FR Easy	1
200 D IM Any Easy	2
100 K Any Easy	3

First set	
4 × ⎰ **200** K IM Order ◔0:15	4
100 FR @_1:45	5
4 × **25** K FL ◔1:00 Race Pace ↧ Fins	6
4 × **200** as ⎰ **50** K Front Flutter	7
50 K Side Flutter	8
50 K Back Flutter	9
50 K Side Flutter	10

Warm down	
200 Any Easy	11

made with: **swiML**

Regular Program 10

Figure 150: Regular program 10.

Christoph Bartneck

This program is a Christmas tradition at the JASI Masters club. December is summer in New Zealand and we swim this in the 50 meter outdoor pool.

⤳**Date:**19 July 2024
⤳**Pool Size:**50
⤳**Units:**meters
⤳**Length:**10000 meters / 200 Laps

100 times 100	
10 × **100** D IM Any @_2:00	1
10 × **100** FR @_1:45	2
10 × **100** FR @_1:45 Pads	3
10 × **100** FR @_1:45 Pullbuoy	4
10 × **100** FR @_1:45 Pads Pullbuoy	5
10 × **100** K IM Overlap ⌚0:15	6
10 × 50…90% **100** FR @_1:45	7
10 × **100** as **50** Nr 2	8
50 Nr 3	9
10 × **100** as **50** Nr 1 Easy	10
50 Nr 1 Race Pace	11
10 × **100** Any @_2:00 Easy	12

made with: **swiML**

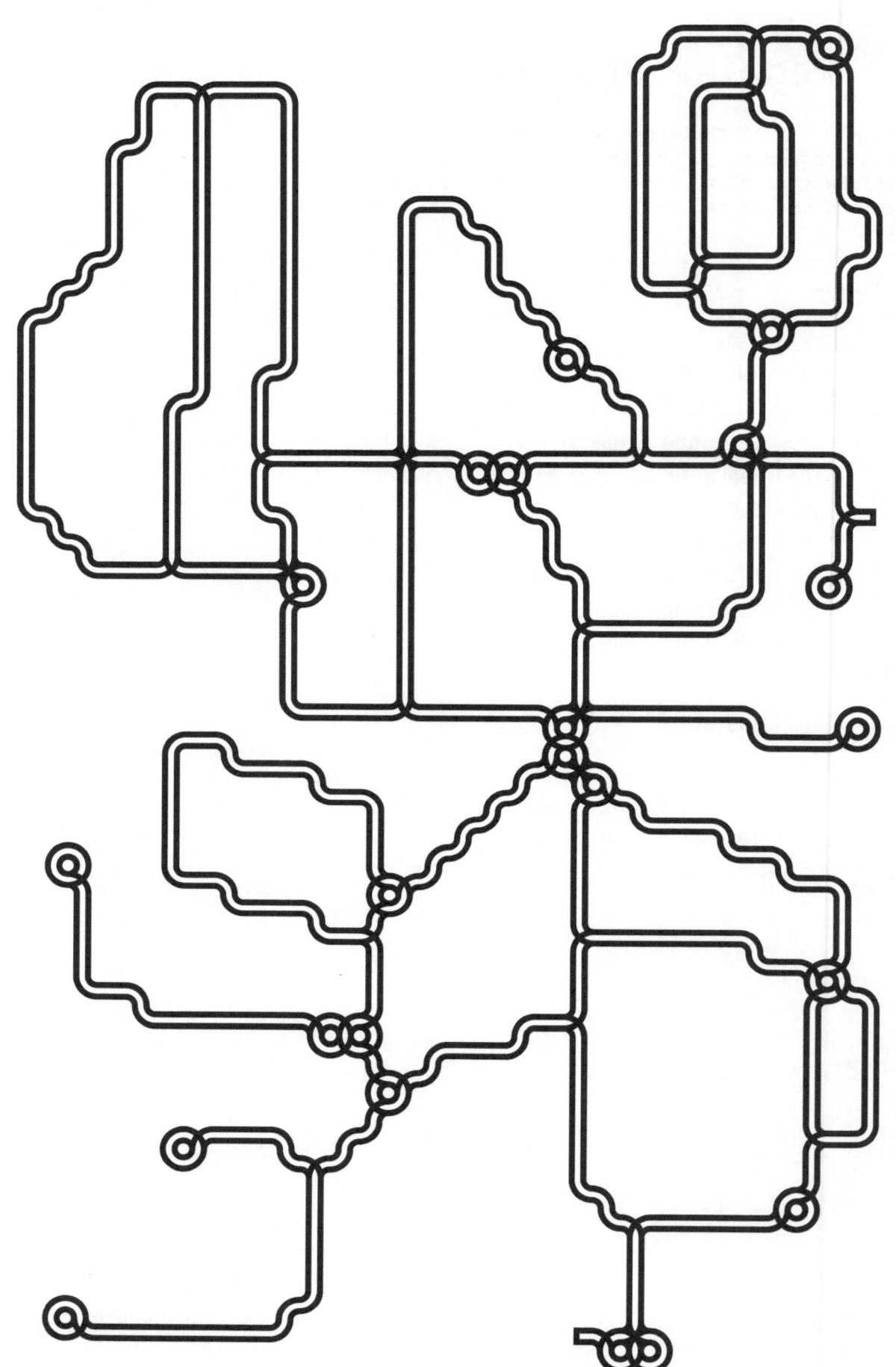

Tools

Summary

This chapter shows how to use software tools to quickly and easily write swiML programs. This includes tools for authoring, validating and transforming training programs. This chapter also demonstrates how to use the Python programming language to author complex training programs.

In this chapter, you will learn how to use software tools to write swiML programs yourself.

Writing swiML

There are many free and commercial editors available that speed up the process of writing, validating and transforming swiML. The underlying XML can be written in any text editor, but modern code editors have many features, such as code highlighting and code completion, that make it that much easier. Here is a list of popular XML editors.

- **Visual Studio:** This full featured Integrated Development Environment (IDE) has not onlyan XML editor but also an XML schema designer and XSLT debugger. Unfortunately, it is only available for Windows. `https://visualstudio.microsoft.com`

- **Visual Studio Code:** This code editor has XML extensions, such as the one from Red Head, that enable many convenient features. `https://code.visualstudio.com`

- **Dreamweaver:** Adobe's Dreamweaver has built in support for editing XML files. `https://www.adobe.com/nz/products/dreamweaver.html`

- **Oxygen Editor:** The Oxygen products offer the most professional tools for all XML needs. `https://www.oxygenxml.com`

Authoring

The first task of an Editor is to speed up the process of writing swiML. One of the most convenient features is auto code completion. Instead of having to write out every XML element, the editor knows the structure of swiML

Figure 151: The Oxygen editor is a professional XML authoring and development tool. (`https://www.oxygenxml.com`)

and offers valid options while you type. In the following paragraphs, we will use the Oxygen Editor (see Figure 151) to showcase how this works. Other editors will have similar functionality. We start by creating a new document by clicking on the "New Document" icon (see 1 in Figure 152) or by using the keyboard shortcut ⌘ + N or the File ⟩ New menu.

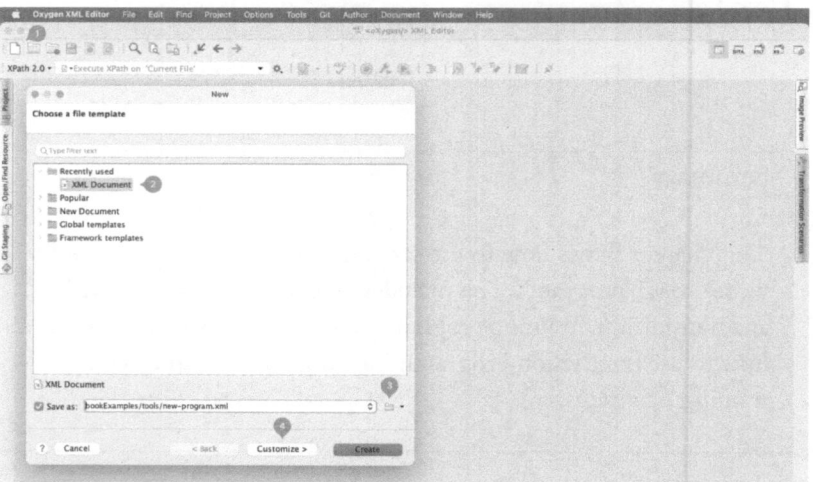

Figure 152: Creating a new document.

In the new document dialogue window, we can then select "XML Document" as the template and choose a location of the new document on our computer (3). Next, we click on the "Customize" button (4) to define the new document further.

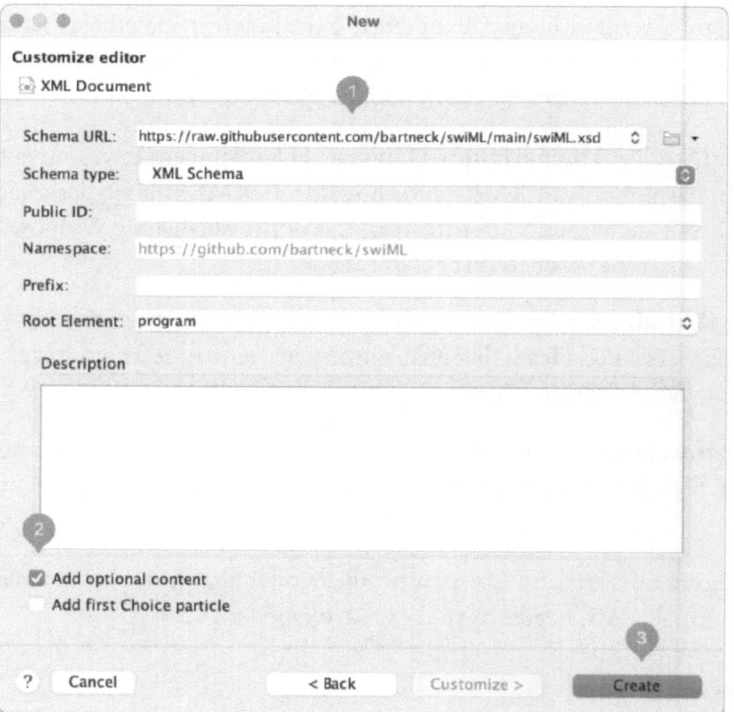

Figure 153: Customising the new document.

First, we enter the URL of the swiML Schema (see 1 in Figure 153). The schema defines the markup language and is stored in an XML Schema

Definition (XSD) file. We store this file for everybody to use on GitHub (see 21). The URL is `https://raw.githubusercontent.com/bartneck` `/swiML/main/swiML.xsd`

Next, we check the option "Add optional content" (2). This will populate the new document with some essential elements defined in the schema. Last, we click on "Create" to return to the previous dialogue (3). There, we click on "Create" again to complete the setup.

We now have a skeleton swiML document (see Figure 154). You can notice that many elements are underlined in red, which indicates that something is wrong. We will discuss validation shortly, but for now, the main issue is that many elements are simply empty.

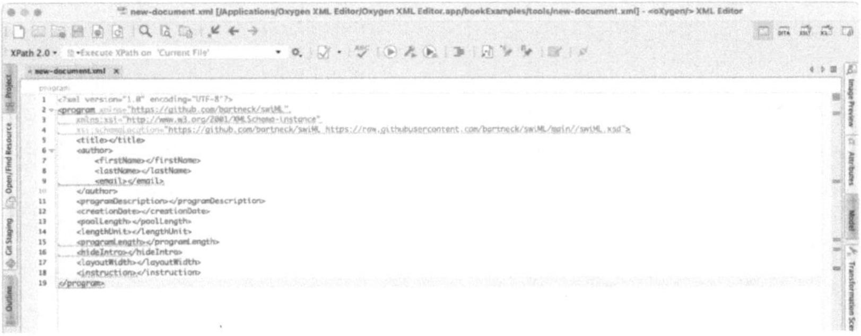

Figure 154: Skeleton swiML document.

We can start filling in missing elements and deleting some we do not need right now (see Listing 32).

```
1   <title>My First Program</title>
2   <author>
3       <firstName>Christoph</firstName>
4       <lastName>Bartneck</lastName>
5       <email>christoph.bartneck@canterbury.ac.nz</email>
6   </author>
7   <programDescription>My first swiML program.</programDescription>
8   <creationDate>2024-04-23</creationDate>
9   <poolLength>25</poolLength>
10  <lengthUnit>meters</lengthUnit>
11  <hideIntro>false</hideIntro>
```

Listing 32: Entering details of a new program.

As a result, we are left with only one validation error. If we hover over the `<program>` element, it informs us that it is expecting an `<instruction>` element. It even offers us a quick fix to insert an `<instruction>` right away (see Figure 155). While this would be convenient, we need to learn how to add new elements in general.

Figure 155: Hovering over an invalid element.

```
program
 1   <?xml version="1.0" encoding="UTF-8"?>
 2 ▽ <program xmlns="https://github.com/bartneck/swiML"
 3
 4                                                         https://raw.
 5       Validation:
 6 ▽        ⓘ ⊞ The content of element 'program' is not complete. One of
 7             '{"https://github.com/bartneck/swiML":layoutWidth, "https://
 8             github.com/bartneck/swiML":instruction}' is expected.
 9             2 quick fix(es) available:
10             <◈> Insert element 'instruction'
11                                          Press F2 for focus  escription>
12       <creationDate>2024-04-23</creationDate>
13       <poolLength>25</poolLength>
14       <lengthUnit>meters</lengthUnit>
15       <hideIntro>false</hideIntro>
16   </program>
```

If we only enter the < symbol, the editor offers us auto-complete options (see Figure 156), including helpful annotations that explain each choice. We select the instruction option by pressing the ⏎enter key.

Figure 156: Using auto-complete to add an <instruction> element.

```
program
 1   <?xml version="1.0" encoding="UTF-8"?>
 2 ▽ <program xmlns="https://github.com/bartneck/swiML"
 3       xmlns:xsi="http://www.w3.org/2001/XMLSchema-instance"
 4       xsi:schemaLocation="https://github.com/bartneck/swiML https://raw.
 5       <title>My First Program</title>
 6 ▽     <author>
 7           <firstName>Christoph</firstName>
 8           <lastName>Bartneck</lastName>
 9           <email>christoph.bartneck@canterbury.ac.nz</email>
10       </author>
11       <programDescription>My first swiML program.</programDescription>
12       <creationDate>2024-04-23</creationDate>
13       <poolLength>25</poolLength>
14       <lengthUnit>meters</lengthUnit>
15       <hideIntro>false</hideIntro>
16
17       <
18   </progr   ● instruction        The basic elements for programs. Each
                ● layoutWidth         instruction defines what to swim.
                ● /program>
                ● !-- -->
                ● ![CDATA[]]>
```

We now have an <instruction> element that is again underlined in red. As we learned in the section, "Essential Instructions" on page 23, each <instruction> element must contain a <length> and <stroke> element. For now, we focus on using the editor's shortcuts to add elements quickly. If we again type the < symbol, the list of options appears (see Figure 157).

Figure 157: By typing the < symbol we open the auto complete options menu.

```
16
17 ▽     <instruction>
18       <                        Number of arm strokes per breath.
19   </instr  ● breath
20   </program>  ● continue
                 ● equipment
                 ● excludeAlign
                 ● instructionDescription
                 ● intensity
                 ● length
```

If we now press the letter ⎵I⎵, the menu is filtered to only options that start with "l" (see Figure 158). In this case, we were looking for the `<length>` element, which we select by pressing the enter button on our keyboard. With only *three key presses*, we were able to add a new element. Using the advanced features of editors dramatically increases the speed at which you can write swiML code.

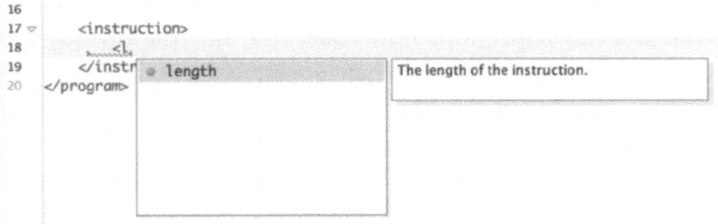

Figure 158: A valid `<instruction>` element.

We continue to use the same keystrokes to add the `<lengthAsDistance>` and `<stroke>` elements and end up with a complete `<instruction>` element that also fulfils the requirement of having at least one `<instruction>` element in the `<program>` element. The `<instruction>` element can have many more optional elements, but for now, we are satisfied with one minimal valid `<instruction>` element (see Figure 159).

```
16
17 ▽     <instruction>
18 ▽         <length>
19              <lengthAsDistance>100</lengthAsDistance>
20          </length>
21 ▽         <stroke>
22              <standardStroke>freestyle</standardStroke>
23          </stroke>
24      </instruction>
```

Figure 159: A valid `<instruction>` element similar to Listing 5 on page 20.

The Oxygen Editor has additional views on our XML document. So far, we only viewed our document in the "Text" view. At the bottom of the window, you find two more: Grid and Author. The latter requires an additional Cascading Style Sheet (CSS), which we do not have for editing swiML documents. But you can use the Grid view, which displays the XML values in a large table (see Figure 160).

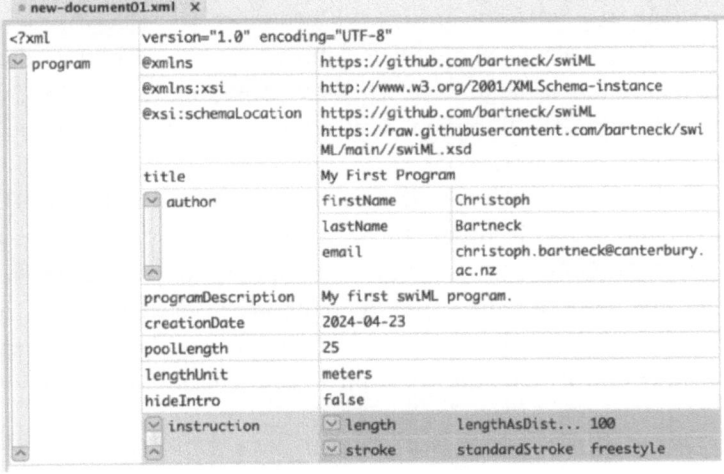

Figure 160: The Grid view on an XML document.

Figure 161: Oxygen is using Apache's Open Source Xerces libraries to validate XML files.

Oxygen is using the Saxon XSLT and XQuery processor.

Validation

Our editor has already indicated several times by underlining elements in red that the document is invalid. This is similar to a grammar checker that you might be familiar with in Microsoft Word or Libre Office. This works because we told the editor at the beginning of the file where the schema is located, as we already learned in the Validation section on page 20. The XML document is continuously validated against the schema. You can also trigger a manual validation by navigating to the menu Document ⟩ ⟩ Validate ⟩ Validate or using the keyboard shortcut cmd + shift + y . Oxygen includes a validation engine that checks an XML file against validation scenarios. Simpler editors might not have validation engines and might even struggle with auto-complete functions that are based on schemas.

Only valid swiML documents can successfully be transformed into HTML documents. The swiML schema includes many assertions to ensure that there are no ambiguities in the program. Having said that, there are some built in uncertainties. If you ask swimmers not to swim freestyle, then they might swim butterfly, backstroke or breaststroke. You cannot be certain which one they will choose. But aside from built in uncertainties, swiML tries to be as precise as possible while giving coaches enough freedom to express their ideas for training programs.

Transformation

Once we have a valid swiML program, we can transform it to other formats, such as HTML, that can be viewed and published online. Oxygen can perform this transformation using Extensible Stylesheet Language Transformations (XSLT). We already discussed this transformation in the Transformation section on page 22. Here, we will learn how to do this using Oxygen. Other code editors might require the installation of extensions or even the whole transformation engine, such as Saxon.

You are also able to start a transformation by using the menu Document ⟩ ⟩ Transformation ⟩ Apply Transformation Scenario(s) or the keyboard shortcut cmd + shift + t . If you run this for the first time, it will bring up a dialogue window to configure the transformation scenarios (see Figure 162). You can later come back to this dialogue by using the menu Document ⟩ Transformation ⟩ ⟩ Configure Transformation Scenario(s) or the keyboard shortcut cmd + shift + c . We need to create a new scenario by clicking on the "New" button that brings up a drop down menu from which we need to select the "XML transformation with XSLT" option. This brings up the "New Scenario" dialogue window (see Figure 163).

Here, we have to give the new scenario a name (1) and add the URL of the XSLT file, which is `https://raw.githubusercontent.com/ba rtneck/swiML/main/swiML.xsl`. In the Output tab we have to set the output filename (see 1 in Figure 164). A common approach is to set it to the current file name plus the extension: `$cfn.html`. It is also convenient to open the HTML file with the operating system's default web browser to see the results (see 2 in Figure 164).

Figure 162: Dialogue window to set up transformation scenario(s).

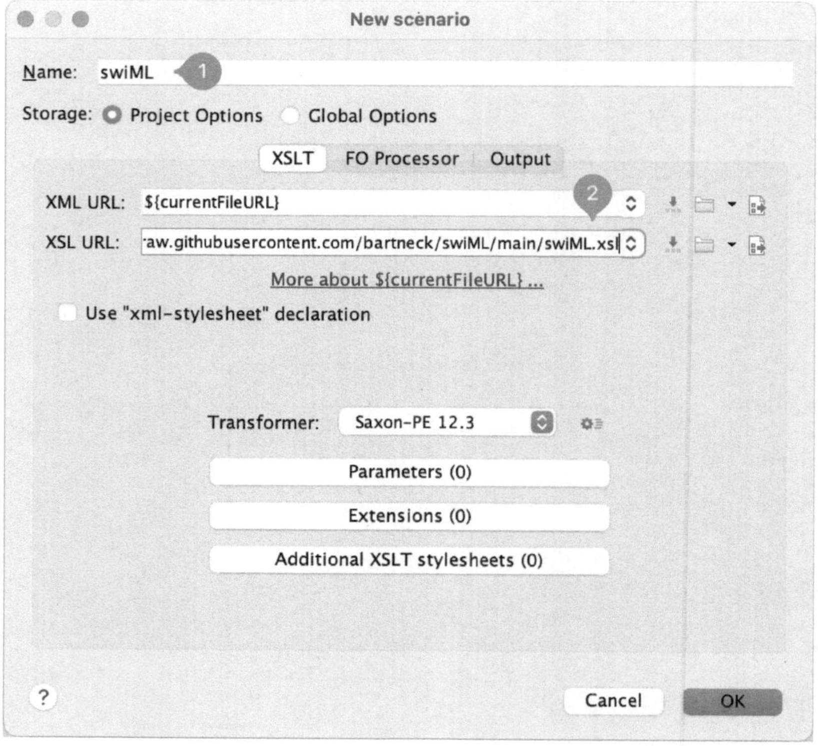

Figure 163: Dialogue window to set up new transformation scenario.

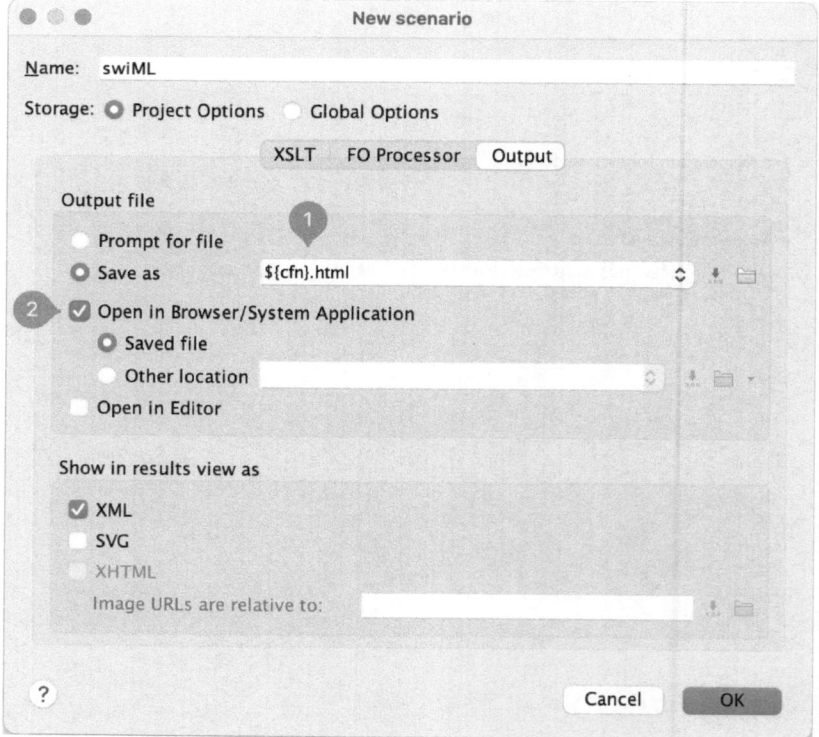

Figure 164: The Output tab of the new scenario dialogue window.

After clicking on "OK" we get back to the New Scenario dialogue, where we can click on the "Apply associated" button to trigger the transformation. This will bring up your web browser with the transformed HTML document (see Figure 165).

Figure 165: Viewing the transformed swiML document in a web browser.

From here, we can upload this HTML file to a web server to share it online. We can also create a PDF document that can be printed and shared. All web browsers can save the current document as PDF through the printing dialogue. You can select File ⟩ Print... or press cmd + p to bring up the printing dialogue. From here you can select "Save as PDF" (2) from the PDF dropdown menu (1).

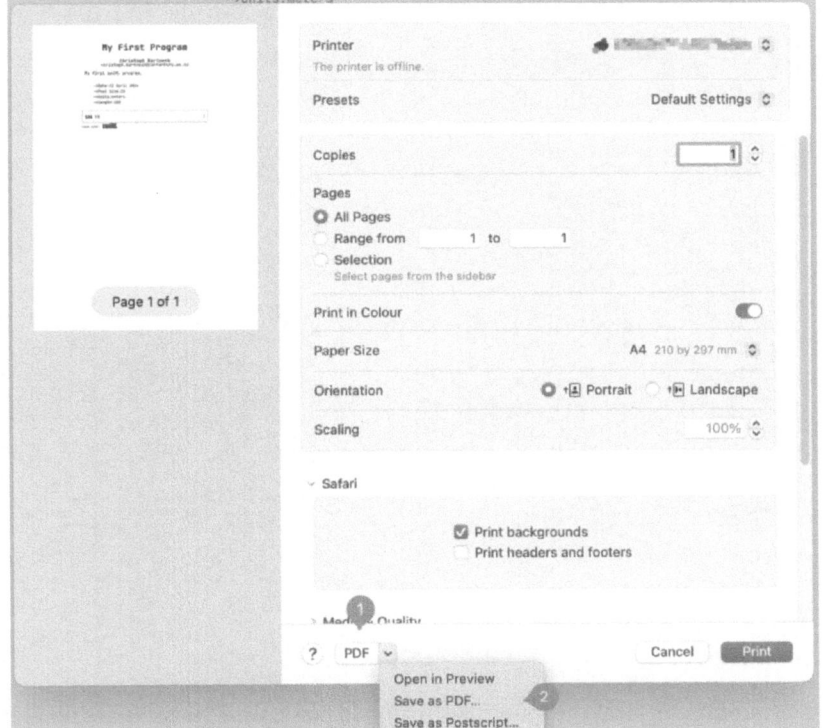

Figure 166: Creating a PDF document from the swiML HTML page.

You can print the PDF and bring it to the pool. Inkjet printers are unsuitable since the ink will dissolve once the paper becomes wet. Some swimmers laminate their printed programs. You can also use a reusable waterproof map case to bring programs to the pool.

Programming

The Python programming language is a powerful tool for generating swiML programs. It is available for all major operating systems, such as Windows, MacOS, and Linux. You can use Python to generate swiML programs.

Online IDE

The easiest way to start with Python and swiML is to use an Integrated Development Environment (IDE). These are even available as an online service. We can program Python directly in our web browser without installing any software on our computer (see Figure 167). The free versions offer basic functionality, but often, a paid subscription is necessary to take full advantage of all its features. Popular examples of such online IDEs are:

- **Replit:** A full featured IDE. The main limitation of the free version is that you cannot keep your code private. `https://replit.com`

- **Visual Studio Code for the Web:** This online version of Microsoft's code editor does not have the same functionality as its locally installed counterpart, but it is good enough to view and edit minor changes. `https://vscode.dev`

Figure 167: The Replit IDE in a web browser.

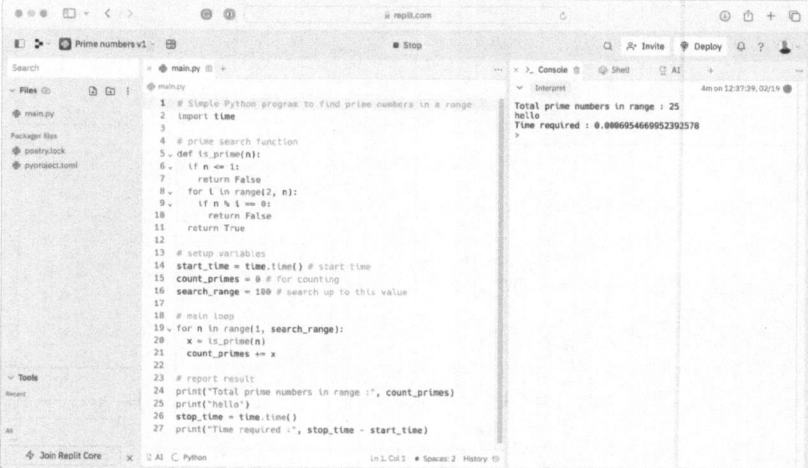

Local IDE

Coding on our local computer has many advantages. The tools are better, they run faster, and we are not dependent on an internet connection. There are a wide range of free IDEs we can use to write Python code. Many of the free IDEs do not even have any limitations. Only complex projects benefit from the convenience of a professional commercial IDE. Here is a list of some free code editors we can use:

- **Sublime:** A popular code editor available as shareware. `https://www.sublimetext.com`

- **Pycharm:** This IDE comes in a free community edition, an educational edition and a professional edition. `https://www.jetbrains.com/pycharm/`

- **Visual Studio Code:** Microsoft's free code editor. Not to be confused with Microsoft Visual Studio, their professional IDE. `https://code.visualstudio.com`

Getting Started

If we decide to use an online IDE, then we do not need to bother with any of the following instructions. It is also not possible to provide instructions for all operating systems. Here, we will focus on MacOS. Windows users can follow the excellent tutorial provided by Microsoft[29]. If you are using Linux, then it can be assumed that you already know how to set up and run Python.

[29] `https://code.visualstudio.com/docs/python/python-tutorial`

Installing Python

It is possible to install Python directly with the installers provided by Python's website[30]. But there are even more convenient tools available.

Package managers simplify the installation and updating of software. Instead of visiting websites to download installers or having each installed software check for updates, a package manager keeps track of all of this centrally. A popular package manager for MacOS is Homebrew[31]. Once it is installed, all other installations will become much easier. While it is possible to install Homebrew using a classical MacOS `.pkg` installer file, it will soon become necessary to become familiar with MacOS's terminal.

The Terminal program is in the `Utilities` folder inside our `Applications` folder. Open it up, and type this code into it. We can also copy and paste it from Homebrew's website.

[30] `https://www.python.org/downloads/`

[31] `https://brew.sh`

```
1   /bin/bash -c "$(curl -fsSL
    ↪  https://raw.githubusercontent.com/Homebrew/install/HEAD/install.sh)"
```

After entering your password, the Homebrew package manager will be installed (see Figure 169).

```
●●●                          Terminal
~ $/bin/bash -c "$(curl -fsSL https://raw.githubusercontent.com/Homebrew/install
/HEAD/install.sh)"
==> Checking for `sudo` access (which may request your password)...
Password:
```

Figure 168: The Terminal is the command line interface to your computer.

Figure 169: Installing the Homebrew package manager using the Terminal.

The next step is to install the latest version of Python. Type this command into your Terminal and hit the ⏎ key:

```
1   brew install python
```

Alternatively, we can download a Python installer directly from Python's website, but the Homebrew installation also installs other useful tools, such as the Pip package manager, which we will cover soon. After the command runs successfully, we need to restart the Terminal for the changes to take effect. Then we can run a simple test to ensure that Python was installed successfully. Enter the command:

```
python3 --version
```

The Terminal will tell us which version of Python is installed. In this case, it was 3.11.7:

Figure 170: Testing the Python installation.

```
~ $python3 --version
Python 3.11.7
~ $
```

[32]https://code.visualstudio.com/docs/setup/mac

The next step is to install an IDE. Microsoft's Visual Studio Code is a good choice, and we can download it directly from Microsoft[32]. But since we are already on the command line, we can simply use Homebrew again:

```
brew install --cask visual-studio-code
```

We are almost done. The last step is to add the Python extension to Visual Studio Code. It enables the editor to better support our coding by code highlighting, code completion and IntelliSense. Open Visual Studio Code, which is located in the Applications folder.

Figure 171: Installing the Python extension in Visual Studio Code.

Then click on the extensions icon in the left bar (1), search for "python" in the search box (2) and click on the install button (3) (see Figure 171).

Now it is time for the obligatory Hello World test program. Write these two lines of code and run it (see Figure 172).

```
msg="Hello World"
print(msg)
```

This program prints "Hello World" to the Terminal within Visual Studio Code. While this program will not win us a Nobel prize, it does test our installation process.

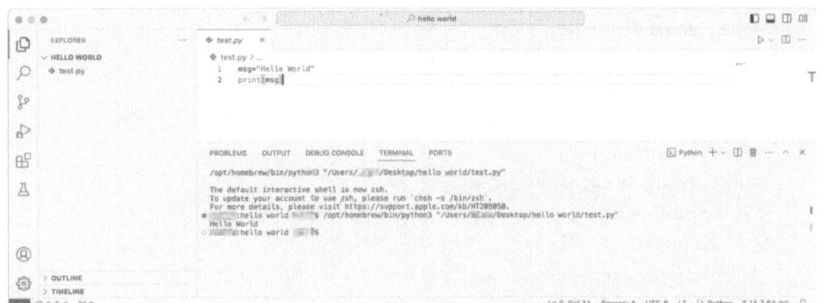

Figure 172: Running the Hello World program in Visual Studio Code.

Now, we can go back to the Pip package manager. It is used to install packages from the Python Package Index (PyPI) (see Figure 173). The swiML packages can be installed directly from there using the following command:

```
pip3 install swiML
```

Figure 173: The Python Package Index is a widely used repository. It hosts more than 500 000 projects. https://pypi.org

You can install this directly in the terminal of your operating system, or you can use the terminal integrated into Visual Studio Code. You might also want to consider using virtual environments for Python, but this will go beyond the scope of this chapter. In any case, you can also simply put the swiML.py file in the same folder as your Python program and then import it in the same way as is shown in Listing 33 on line 1.

The complexity of the tool chain introduced here is above the desirable level. Ideally, just one package manager would install and update all software. Instead, we have to deal with three installation managers: Homebrew, Visual Studio Code Extensions and Pip. The online IDEs do many of these tasks for you and might be a better tool for a beginner.

Writing swiML in Python

Now that we have all the tools in place, we can write our first swiML program in Python (see Listing 33). We start in line 1 with importing swiML. In line 2 we create the myProgram list that will contain all the instructions. Lines 5–9 create a simple warmUp instruction which is added to myProgram in line 11. In the write_program function, many parameters of the program are set and written to a file in line 28.

```python
import swiML
myProgram=[]


# simple instruction
warmUp=swiML.Instruction(
    length=('lengthAsDistance',100),
    stroke=('standardStroke','freestyle'),
    intensity=('startIntensity',('zone','easy')),
)

```

```
11   myProgram.append(warmUp)
12
13   #writing the swiML program to disk
14   def write_program():
15       # create the program
16       simpleProgram=swiML.Program(
17           title='My First Python swiML Program',
18           author=[('firstName','Christoph'),('lastName','Bartneck')],
19           programDescription='This is a small program to demonstrate how
             ↪  to create a swiML program in Python.',
20           poolLength='25',
21           creationDate='2024-05-20',
22           lengthUnit='meters',
23           swiMLVersion='2.1',
24           hideIntro=False,
25           instructions=myProgram
26       )
27       # write swiML XML to file
28       swiML.writeXML('my_first_swiML_program.xml',simpleProgram)
29
30   # write the swiML program
31   write_program()
```

Listing 33: A first swiML program written in Python.

When you run the program, it creates a new swiML XML file called my_first_swiML_program.xml. You can inspect it directly in Visual Studio Code, or you can open it in Oxygen.

```
1   <?xml version='1.0' encoding='utf-8'?>
2   <program xmlns="https://github.com/bartneck/swiML"
    ↪  xmlns:xsi="http://www.w3.org/2001/XMLSchema-instance"
    ↪  xsi:schemaLocation="https://github.com/bartneck/swiML/version/2/2.1
    ↪  https://raw.githubusercontent.com/bartneck/swiML/main/version/2/⌋
    ↪  2.1/swiML.xsd">
3     <title>My First Python swiML Program</title>
4     <author>
5       <firstName>Christoph</firstName>
6       <lastName>Bartneck</lastName>
7     </author>
8     <programDescription>This is a small program to demonstrate how to
      ↪  create a swiML program in Python.</programDescription>
9     <creationDate>2024-05-20</creationDate>
10    <poolLength>25</poolLength>
11    <lengthUnit>meters</lengthUnit>
12    <hideIntro>false</hideIntro>
13    <instruction>
14      <length>
15        <lengthAsDistance>100</lengthAsDistance>
16      </length>
17      <stroke>
18        <standardStroke>freestyle</standardStroke>
19      </stroke>
20      <intensity>
21        <startIntensity>
22          <zone>easy</zone>
```

```
23        </startIntensity>
24       </intensity>
25     </instruction>
26   </program>
```

Listing 34: Entering details of a new program.

In Oxygen, you can trigger the transformation to HTML, which will generate the expected visual representation of the program (see Figure 174).

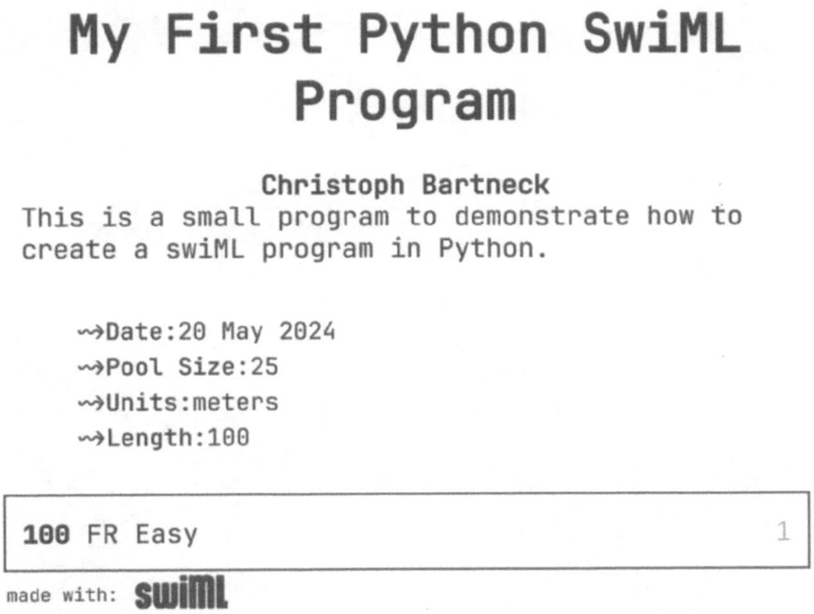

My First Python SwiML Program

Christoph Bartneck
This is a small program to demonstrate how to create a swiML program in Python.

⤳Date:20 May 2024
⤳Pool Size:25
⤳Units:meters
⤳Length:100

| **100** FR Easy | 1 |

made with: **swiML**

Figure 174: The visual representation of the my_first_swiML_program program.

Now that you have access to the full power of a programming language, you can use its features to create intriguing swim training patterns. A first step could be to use Python loops, such as in the Loops section on page 172. The pattern's concept is too complex for expressing it with swiML's `<repetition>` feature.

Conclusions

You now have the tools not only to write your own swiML programs directly in a convenient XML editor, but also to create far more complex programs using the Python programming language. A swimming program is fundamentally a simple programming language, and using the libraries and tools provided by Python empowers you to create sophisticated programs that will keep the swimmers in your lane challenged.

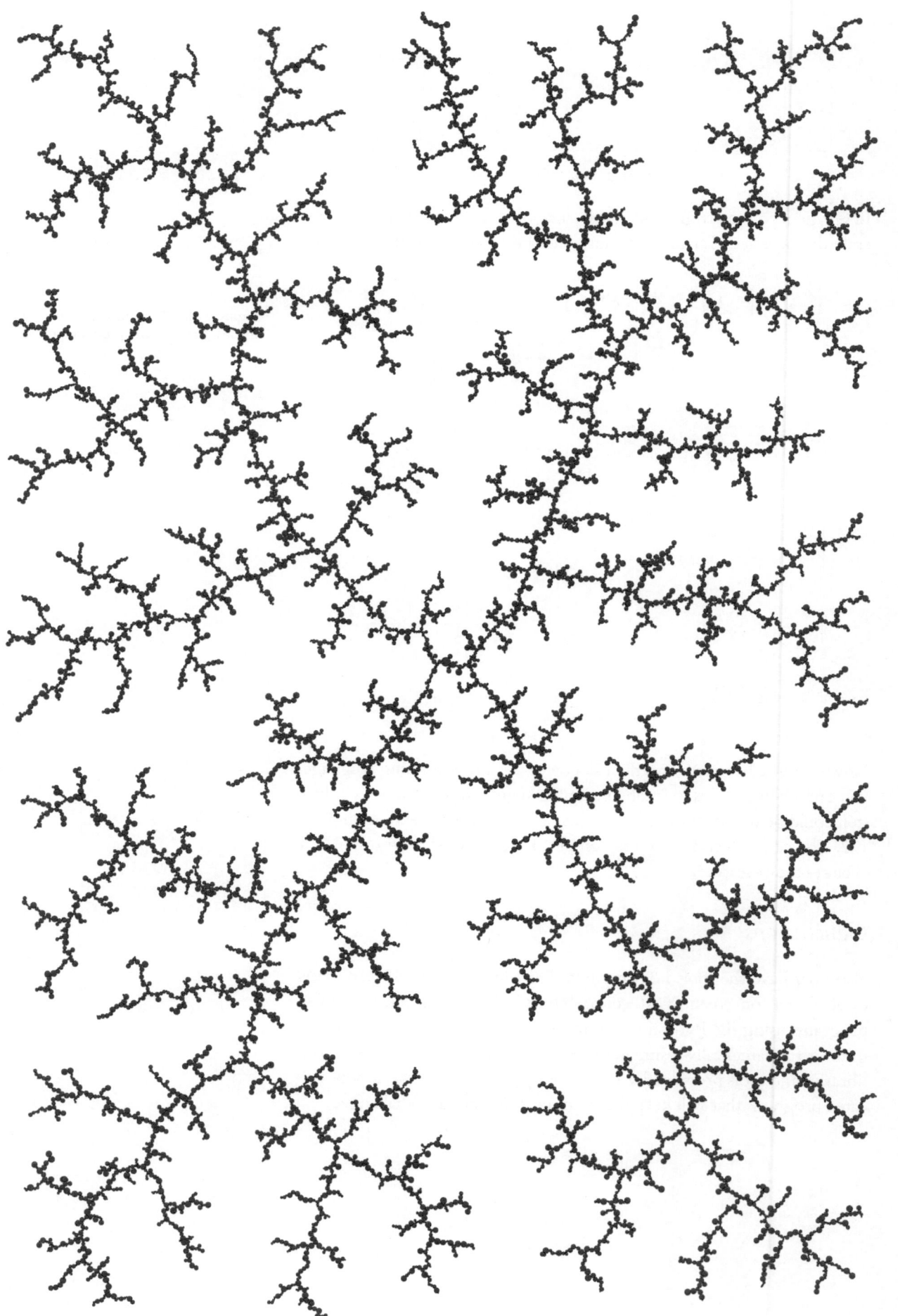

swiML Schema Reference

Summary

This chapter provides an in-depth reference to the structure and elements of the swiML language. Examples provide insights into how to apply the elements. In addition, illustrated descriptions of the strokes and drills are made available.

The XML Schema Definition (XSD) schema defines what is valid swiML. The schema is specified in the `swiML.xsd` file. You can validate your swiML program against this schema to ensure that it will be suitable for further processing, such as transforming it to HTML. The complete documentation of the schema is available in our GitHub repository[33].

[33] https://swiml.org/documentation/XSDdocs/swiML.html

The sequence of XML elements does, in some cases, matter. For example, for an ordinary `<instruction>` element, the first nested element has to be `<length>` followed by `<stroke>`. For other elements, the sequence does not matter. For example, it does not matter if you write:

```
1   <equipment>pads</equipment>
2   <equipment>pullBuoy</equipment>
```

Listing 35: Sequence of equipment 1.

or:

```
1   <equipment>pullBuoy</equipment>
2   <equipment>pads</equipment>
```

Listing 36: Sequence of equipment 2.

Meta

The meta information elements provide information about the program itself. This allows you to describe the program and make it searchable. A variety of elements is available:

<title>

The title of the program is limited to a length of 60 characters.

<author>

The author of the program. This element requires a firstName and last-Name. Adding an email address is optional.

<programDescription>

A description of the program. Its length is limited to 400 characters.

<creationDate>

The date of the creation of the program. Dates are specified as dd-mm-yyyy. For example, 24-12-2025.

<poolLength>

The length of the pool for which the program is designed. This is necessary if swimming instructions are given in laps. While any swim program defined for a 50-meter pool can be swum in a 25-meter pool, the reverse cannot be guaranteed.

<lengthUnit>

The unit of measurement for the poolLength. It can be set to meters or yards.

<programAlign>

This element can be set to true or false. When set to false, then no <instructions> will be horizontally aligned. This element overwrites the specific align elements inside instructions.

<numeralSystem>

This element will set the numeral system to either Roman or the default Decimal.

<hideIntro>

This element can be set to true or false. When true, it hides all the information in the header of the program, such as its title, author, description, etc.

<layoutWidth>

The width of rendering the program can be set to fit different formats. The unit used is characters, where the font size is set to 3.67 mm. The <layoutWidth>, therefore, uses a horizontal integer measurement while the width of each letter is defined by a vertical measurement in mm. Since swiML uses a monospaced typeface, all characters have the exact same width. 50 characters fit exactly into an 11 cm wide box. Each letter is therefore $\frac{110}{50} = 2.2$ mm wide and 3.67 mm height. For an A4 paper print, 80 characters are suitable.

The question of what exactly the size of a typeface is requires an explanation. One popular misconception is that it refers to the highest ascender to the lowest descender of the letters. In the example shown in Figure 175, this would be from the highest point of the letter M to the lowest point of the letter y.

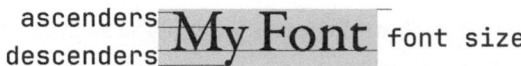

Figure 175: The baseline, ascender, descender and font height.

In reality, the font size refers to more than that. There is additional space above and below. This space is based on the emerging movable type printing technology around 1450 by Johannes Gutenberg. Each individual letter was made from a block of metal that would be assembled to create words, sentences and paragraphs (see Figure 175).

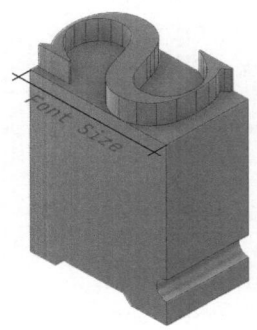

Figure 176: A metal sort for the letter S.

The vertical size of the block of metal determines the size of the font. There were several competing standards for the unit of measurement for the height. They even confusingly shared the same name: "Point". There were French, German, Japanese, British and American Points. Possibly due to the dominance of American Desktop Publishing (DTP) software, the American point, defined as $\frac{1}{72}$ of an inch, became the de facto standard. One American point is approximately $0.352\overline{7}$ mm. If you select a 10-point Arial in Microsoft Word, then your typeface will fit into a box that is around 3.5 mm tall. swiML does not use the traditional point units. Instead, we use the more easy to understand and manipulate unit of millimetres.

<program>

The <program> element is the root element. It includes attributes on the name space, the xsi name space prefix and its schema location. A typical example is shown in Listing 37.

```
1  <?xml version="1.0" encoding="UTF-8"?>
2  <program xmlns="https://github.com/bartneck/swiML"
3      xmlns:xsi="http://www.w3.org/2001/XMLSchema-instance"
```

```
4    xsi:schemaLocation="https://github.com/bartneck/swiML/swiML
     ↪   https://raw.githubusercontent.com/bartneck/swiML/main/
     ↪   swiML.xsd">
```

Listing 37: The `<program>` element with all its attributes.

Versions   To ensure backward compatibility of the swiML schema, we developed a versioning system based on the `xsi:schemaLocation` attribute. The most recent schema currently under development is available in the root directory as shown in Listing 37. Stable releases are published in the versions folders as shown for version 2.1 in Listing 38. The latest stable version is published in the `versions/latest/` folder.

```
1    <?xml version="1.0" encoding="UTF-8"?>
2    <program xmlns="https://github.com/bartneck/swiML"
3        xmlns:xsi="http://www.w3.org/2001/XMLSchema-instance"
4        xsi:schemaLocation="https://github.com/bartneck/swiML/version/2/2.1
         ↪   https://raw.githubusercontent.com/bartneck/swiML/main/version/
         ↪   2/2.1/swiML.xsd">
```

Listing 38: The `<program>` element for the 2.1 version.

Meta Example

We can now put all the meta elements together for a minimal swiML program (see Listing 39). The sequence of the meta elements matters.

```
1    <?xml version="1.0" encoding="UTF-8"?>
2    <program xmlns="https://github.com/bartneck/swiML"
3        xmlns:xsi="http://www.w3.org/2001/XMLSchema-instance"
4        xsi:schemaLocation="https://github.com/bartneck/swiML https://
         ↪   raw.githubusercontent.com/bartneck/swiML/main/swiML.xsd">
5        <title>Example Program</title>
6        <author>
7            <firstName>Christoph</firstName>
8            <lastName>Bartneck</lastName>
9            <email>christoph@bartneck.de</email>
10       </author>
11       <programDescription>This is a minimal program.</programDescription>
12       <creationDate>2024-04-22</creationDate>
13       <poolLength>25</poolLength>
14       <lengthUnit>meters</lengthUnit>
15       <programAlign>true</programAlign>
16       <hideIntro>false</hideIntro>
17       <layoutWidth>50</layoutWidth>
18
19       <instruction>
20           <length>
21               <lengthAsDistance>100</lengthAsDistance>
22           </length>
23           <stroke>
24               <standardStroke>freestyle</standardStroke>
```

```
25        </stroke>
26      </instruction>
27    </program>
```

Listing 39: A minimal program with all the meta elements.

Instruction

The <instruction> element is the key to all programs. It can either be used to define a structure or a direction. A structural element can be a <repetition>, <continue> or <segmentName>. The first two elements can be nested. A direction provides information on what exactly to swim next. Let us start with the structural elements.

<repetition>

This is probably one of the most frequently used types. It allows the author to define that the nested <instruction> elements are to be repeated. A repetition is visualised using the [symbol.

The number of repetitions is set with the <repetitionCount> element. We already encountered a simple repetition in Listing 9 on page 25 and a nested repetition in Listing 10 on page 26. Here, we will only discuss the extended functionality.

Compact   Often, repetitions only vary on one aspect. For example, a repetition might completely use the freestyle stroke and only vary the equipment. The <stroke> element and all other options for the <instruction> element can be moved directly into the <repetition> element. Listing 40 shows how the <stroke> element is moved directly into the <repetition> element (lines 4-6).

swiML includes an assertion in its schema to ensure that no conflicting information can be written. For example, if a <stroke> element is specified directly in the <repetition> element, then no <stroke> element is allowed in any of the nested instructions.

Simplify   To make the swiML code even more concise, it includes the option of simplifying repetitions using the <simplify> element, which can be set to true or false. By default, it is set to false. If it is set to true, then the behaviour and rendering of the <repetition> element changes. It is no longer necessary to specify a <repetitionCount> since it is automatically calculated from the nested instructions. For this to work, all nested instructions must have the same length. Nested <continue> elements must also add up to exactly the same length and must be declared using the <continueLength> element.

The <repetitionCount> of nested repetitions becomes a counter. It is no longer a multiplier. The visualisation also changes to that listed in Listing 40 and shown in Figure 177.

```
1    <instruction>
2        <repetition>
3            <simplify>true</simplify>
4            <stroke>
5                <standardStroke>freestyle</standardStroke>
6            </stroke>
7            <instruction>
8                <repetition>
9                    <repetitionCount>2</repetitionCount>
10                   <instruction>
11                       <length>
12                           <lengthAsDistance>100</lengthAsDistance>
13                       </length>
14                       <equipment>fins</equipment>
15                   </instruction>
16               </repetition>
17           </instruction>
18           <instruction>
19               <length>
20                   <lengthAsDistance>100</lengthAsDistance>
21               </length>
22               <equipment>pullBuoy</equipment>
23           </instruction>
24       </repetition>
25   </instruction>
```

Listing 40: A simplified and compacted repetition.

Notice how the number of repetitions declared in Listing 40 are used as counters. The total number of repetitions is calculated automatically (see Figure 177).

Figure 177: In a simplified repetition the <repetitionCount> elements become a counters.

Alignment   swiML will try to vertically align elements in the HTML layout. <repetition> elements will only align to other <repetition> elements within the same segment (see Figure 178). The <excludeAlign> element is used to exclude the repetition from the alignment. The element is the last in the sequence of an <instruction>.

Figure 178: Two aligned repetitions.

There is a threshold for the alignment process. If two elements are more than 10 characters apart, then they will not be aligned (see Figure 179). In addition, the `<excludeAlign>` element explicitly excludes an element from the layout process.

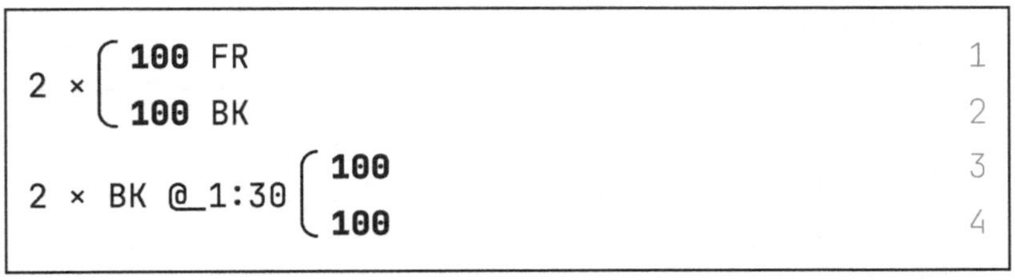

Figure 179: These two repetitions are too far apart to be aligned.

\<continue\>

This element allows the author to define that the nested `<instruction>` elements are to be swum without any rest between them. A continue element is visualised using the | symbol. We already discussed the basic functionality in the Continue section on 28.

\<continueLength\>   The `<continueLength>` element adds an aspect of a repetition to the `<continue>` construct. Let's consider the example of the swiML code shown in Listing 41. This is a typical usage scenario for the `<continueLength>` element.

```
1   <instruction>
2       <continue>
3           <continueLength>
4               <lengthAsDistance>200</lengthAsDistance>
5           </continueLength>
6           <instruction>
7               <length>
8                   <lengthAsDistance>50</lengthAsDistance>
9               </length>
10              <stroke>
11                  <standardStroke>freestyle</standardStroke>
12              </stroke>
13          </instruction>
14          <instruction>
15              <length>
16                  <lengthAsDistance>50</lengthAsDistance>
17              </length>
18              <stroke>
19                  <standardStroke>backstroke</standardStroke>
20              </stroke>
21          </instruction>
22      </continue>
23  </instruction>
```

Listing 41: The `<continueLength>` element in use.

When the <continueLength> element is excluded, then this means swimming 50 freestyle followed by 50 backstroke (see lines 1-2 in Figure 180). This adds up to 100 meters. If the <continueLength> element is included, in this example, it is set to 200, then the 50 freestyle and 50 backstroke need to be repeated twice (see lines 3-4 in Figure 180), adding up to 200 meters of swimming.

```
1          | 50 FR
   100 as  |
2          | 50 BK
3          | 50 FR
   200 as  |
4          | 50 BK
```

Figure 180: The usage of the <continueLength> element.

Alignment swiML will try to vertically align elements in the HTML layout. <continue> elements will only align to other <continue> elements within the same segment (see Figure 178). The <excludeAlign> element is used to exclude the repetition from the alignment (see Figure 182). The element is the last in the sequence of an <instruction>.

<segmentName>

This element can be used to define segments in the program, such as "Warm up", "Main set" or "Warm down". The names can be freely written by the author; there are no pre-defined names. We already discussed the basic functionality in the Segments section on page 27.

Directions

Instructions that direct the swimmer on what to do next are the basic building blocks of all programs. There are a wide range of options available. Some of the elements need to occur in a sequence. These are:

1. Length
2. Stroke
3. Rest
4. Intensity
5. All other elements

We will now define the elements above in more detail.

Length Each <instruction> element must have a <length> element. Within the <length> element, one of the three types of length are expected. These are:

<lengthAsDistance> The length expressed as a unit of length. The unit of measurement is defined in the <lengthUnit> element, which can be set to meters or yards.

<lengthAsLaps> The length expressed in the number of laps. Since the length of the pool is defined in the <poolLength> element, the distance swum can be converted to meters and vice versa.

<lengthAsTime> The length expressed as a period using the ISO format described on page 24. An example would be PT1M30S, which is equivalent to 1:30 minutes.

<stroke>   Each <instruction> element needs to have one <stroke> element. The stroke can be of three types: standardStroke, kicking and drill. The standardStroke type consists of all the major strokes, such as butterfly, backstroke, breaststroke and freestyle. It also offers variations, such as not freestyle, any, or number one. Here is a full list:

- butterfly
- backstroke
- breaststroke
- freestyle
- individualMedley
- reverseIndividualMedley
- individualMedleyOverlap
- individualMedleyOrder
- reverseIndividualMedleyOrder

- any
- nr1
- nr2
- nr3
- nr4
- notButterfly
- notBackstroke
- notBreaststroke
- notFreestyle

The nr1 to nr4 denotes the preference of the swimmer. If he is a breaststroker then his nr1 is breaststroke. notBreaststroke and its variations denote that the swimmer can swim anything but breaststroke. Individual medley has several alternatives that require a closer look.

individualMedley the <lengthAsDistance> is divided into four equal parts. Swim the parts in the medley sequence without a rest between them. Here is an example: Swim 100 Individual Medley as 25 butterfly, 25 backstroke, 25 breaststroke and 25 freestyle without any rest between them.

reverseIndividualMedley similar to individualMedley, but in reversed order.

individualMedleyOrder This stroke only works within a repetition. 4 × 100 is swum as 100 butterfly, 100 backstroke, 100 breaststroke and 100 freestyle.

indivdualMedleyOverlap This stroke only works within a repetition. Each <lengthAsDistance> of an instruction is divided into two parts. Swim both parts without a rest between them. Swim the two parts in the individualMedley order. For a repetition divisible by four, such as 4 × 100, this results in 50 butterfly/50 backstroke, 50 backstroke/50 breaststroke, 50 breaststroke/50 freestyle, 50 freestyle/50 butterfly. For a repetition divisible by three, such as 3 × 100, the last repetition, which is normally freestyle/butterfly, is excluded.

The stroke can also be defined as kicking by using the <kicking> element. This element must include one of two types. Either a <standardKick> that uses the same list as the <standardStroke> list or a more specific definition. The latter is done by defining an <orientation> and a <legMovement>. The leg movement can be set to flutter, dolphin or scissor. The orientation can be set to:

- front
- back
- left
- right
- side
- vertical

Listing 42 shows an example of how to use the <kicking> element with an <orientation> and a <legMovement> element.

```
 1  <instruction>
 2      <length>
 3          <lengthAsDistance>100</lengthAsDistance>
 4      </length>
 5      <stroke>
 6          <kicking>
 7              <orientation>side</orientation>
 8              <legMovement>flutter</legMovement>
 9          </kicking>
10      </stroke>
11  </instruction>
```

Listing 42: The <kicking> element in use.

The third option for defining the <stroke> is by using the <drill> element. This drill element requires two further elements: <drillName> and drillStroke. The latter again uses the <standardStroke> list. There is a long list of possible drills that can be combined with the <drillStroke. Here is the list of the currently included drills:

- 6KickDrill
- 8KickDrill
- 10KickDrill
- 12KickDrill
- fingerTrails
- fist
- 123
- bigDog
- scull
- singleArm
- any
- technic
- dogPaddle
- tarzan
- 3Kick1Pull
- other

Undoubtedly, there will be many more drills, and we will continue to expand the list. You can use the instructionDescription element to describe any missing drills. An explanation of how all the listed drills work is available in the "Drills" section on page 223. Listing 43 shows how the <drill> element can be used.

```
1   <instruction>
2       <length>
3           <lengthAsDistance>100</lengthAsDistance>
4       </length>
5       <stroke>
6           <drill>
7               <drillName>fingerTrails</drillName>
8               <drillStroke>freestyle</drillStroke>
9           </drill>
10      </stroke>
11  </instruction>
```

Listing 43: The <drill> element in use.

<rest> The rest immediately follows a direction to swim. If the rest is given as time, then it follows the ISO format described on page 24. The options are:

<afterStop> Duration of rest after stopping a swimming instruction. Example: 20 seconds means that the swimmer will rest for 20 seconds after stopping the current instruction. This would be expressed as PT0M20S.

<sinceStart> The interval on which swimming instructions start. Example: on 1:30 means that the next instruction starts after 1:30 from starting the current instruction. This would be expressed as PT1M30S.

<sinceLastRest> The time since the end of the last rest. This is useful when several instructions without a rest period are swum, followed by a <sinceStart> type rest.

<inOut> Number of swimmers arriving. Example: 3rd in 1st out: Once the 3rd swimmer in the lane arrives, the 1st swimmer starts.

<intensity> The intensity defines the effort level. This will vary between swimmers and can be expressed in three ways:

<precentageEffort> Effort in percentage. Example: 100 means maximum effort.

<zone> There are five zones available: easy, threshold, endurance, race pace and max.

<precentageHeartRate> Heart rate as a percentage of maximum heart rate.

If the intensity should change within an <instruction>, then another <stopIntensity> needs to be added. This will define the intensity at the start and end of the <instruction>.

```
1   <instruction>
2       <length>
3           <lengthAsDistance>100</lengthAsDistance>
```

```
4      </length>
5      <stroke>
6          <standardStroke>freestyle</standardStroke>
7      </stroke>
8      <intensity>
9          <startIntensity>
10             <zone>easy</zone>
11         </startIntensity>
12         <stopIntensity>
13             <zone>max</zone>
14         </stopIntensity>
15     </intensity>
16  </instruction>
```

Listing 44: A swiML program with a dynamic intensity.

This instruction would be rendered in HTML as shown in Figure 181:

100 FR Easy…Max   1

Figure 181: This HTML render shows the
<instruction> defined in Listing 44.

The intensity can be changed not only within an <instruction> but
also across. It is common to swim four times 100 meters freestyle with
the intensity changing from the first 100 to the last 100. To accom-
plish this, the <intensity> element has to be moved directly under the
<repetition> element. The solution is to add the <startIntensity> and
<stopIntensity> to the <repetition> element (see Listing 45)

```
1   <instruction>
2       <repetition>
3           <repetitionCount>4</repetitionCount>
4           <intensity>
5               <startIntensity>
6                   <zone>easy</zone>
7               </startIntensity>
8               <stopIntensity>
9                   <zone>max</zone>
10              </stopIntensity>
11          </intensity>
12          <instruction>
13              <length>
14                  <lengthAsDistance>100</lengthAsDistance>
15              </length>
16              <stroke>
17                  <standardStroke>freestyle</standardStroke>
18              </stroke>
19          </instruction>
20      </repetition>
21  </instruction>
```

Listing 45: A swiML program with an intensity that changes across the repetition.

> **Bonus**
>
> There is no standard for definitions and nomenclatures for training intensity zones. The "easy" zone is also known as "endurance" or "recovery". Even the number of zones is a matter of debate. The easier models typically have only three zones: easy, medium, and hard. More complex models use nine zones. Defining the zones has to take into account the metabolism of the athlete. Heart rate, breathing, lactic acid production, and the source of energy (fat or carbohydrate) all play a role. It would go beyond the scope of this book to try to define the zones precisely or to list all the synonyms for the zones' names.

<equipment>   The optional `<equipment>` element allows you to specify what types of equipment are going to be used. The available types are:

- board
- pads
- pullBuoy
- fins

- snorkel
- chute
- stretchCord

<breath>   The `<breath>` element allows you to define the strokes per breath. Setting `<breath>` to 3 means that the swimmer is supposed to breathe on every third stroke.

<underwater>   Swimmers spend most of their time on the surface of the water. Swimmers are only allowed to swim the first 15 meters underwater after starts and turns. Still, for training purposes, the coach might suggest swimming underwater. swiML contains the simple `<underwater>` element as part of the `<instruction>` element. It can be set to either `true` or `false`.

<description>   The `<instructionDescription>` element allows you to freely describe further instructions that have not yet been formalised in swiML. It is limited to 100 characters.

<excludeAlign>   swiML will try to horizontally align the `<instruction>` elements in the HTML layout. The scope is limited to `<segmentName>`, meaning that only elements within a segment align with each other. Furthermore, elements will only align with elements of the same type. Repetitions will only align with other repetitions, and a continue will only align with other continues. You may want to exclude some elements from the alignment process, in particular if your program is complex. The `<excludeAlign>` tells the XSLT to exclude this element from the layout process (see Figure 182). The `<excludeAlign>` element is the last in the sequence of an `<instruction>`.

Figure 182: Using exclusion to prevent horizontal alignment.

Instructions	
100 FR	1
1000 FR	2
Repetitions	
100 as **50** FR	3
50 BK	4
100 b5 as **50** FR	5
50 BR	6
Continue	
4 × **50** FR	7
4 × b7 **50** FR	8

Strokes

World Aquatics, formerly known as Fédération Internationale de Natation (FINA, see Figure 183), defines the rules for official swimming competitions. While their rules name strokes and define some limitations, the exact execution of the strokes remains the choice of the swimmer. Here, we can only provide a rough guide to the swimming strokes, so athletes should try to seek feedback and instructions from the coaches to improve their technique.

Figure 183: In 2022 FINA re-branded itself to World Aquatics. FINA was founded in 1908 in London during the Summer Olympics by the Belgian, British, Danish, Finnish, French, German, Hungarian and Swedish swimming federations.

Butterfly

Butterfly is one of the hardest strokes. Both arms must be brought forward simultaneously over the water and brought backward simultaneously under the water. The same holds true for the leg movements. All up and down movements of the legs must be simultaneous. Its four phases are:

1. Entry: During the entry phase, the swimmer extends their arms forward and enters the water with both hands simultaneously. The hands should be slightly angled outward to create an effective entry and minimize resistance. The swimmer simultaneously performs a dolphin kick, where the legs undulate up and down together.

2. Pull: In the pull phase, the swimmer begins to pull their hands backward and downward through the water in a semicircular motion. The elbows bend as the hands move backward, generating propulsion and lifting the upper body out of the water.

3. Push: After the pull phase, the swimmer transitions into the push phase. In this phase, the swimmer quickly pushes their hands and forearms backward, pushing the water forcefully to propel the body forward. The swimmer simultaneously performs a dolphin kick, where the legs undulate up and down together.

4. Recovery: During the recovery phase, the arms are lifted out of the water in a forward motion, similar to the recovery phase in freestyle swimming.

Backstroke

Backstroke is the only stroke that is performed on the back. Your head should stay still looking upwards. The arm movements consist of a continuous, alternating motion between the left and right arms:

1. Recovery: One arm exits the water thumb-first, close to the hip. The arm lifts straight up and over the shoulder in a circular motion, staying relaxed and close to the body. The little finger enters the water first, just past the shoulder line.

2. Pull: After the arm enters the water, the hand turns so that the palm faces outward and downward to catch the water. The arm pulls in a semi-circular motion down and towards the hip, accelerating through the pull. The hand moves from a high position at the catch to a low position at the hip. The hand exits the water near the hip, transitioning smoothly into the recovery phase.

The legs use the flutter kick with one leg kicking upward while the other kicks downward. The kicks should be small and fast, originating from the hips with minimal knee bend. The feet should not break the water's surface too much, and the legs should not drop too deeply.

Breaststroke

Breaststroke is one of the most technical strokes. The arm and leg movements need to be closely coordinated. Let's start with the arm stroke:

1. Outsweep: From a streamlined position with arms fully extended forward, the swimmer begins the stroke by turning the palms outward. The hands move outward and slightly downward, catching the water to prepare for the pull.

2. Insweep: The hands sweep inwards in a semicircular motion, moving towards the chest. The elbows stay high and close to the surface. The swimmer accelerates the hands as they move closer to the midline of the body, providing strong propulsion. The hands end up close to each other under the chest, with elbows tucked in.

3. Recovery: The hands move forward from the chest, sliding just below the surface of the water. The elbows drop to reduce resistance as the hands extend forward. The arms return to the streamlined position, fully extended in front of the body, ready for the next outsweep.

The leg movement is often described as scissor kick or frog kick. It consists of the following phases:

1. Recovery: From the streamlined position, the heels are drawn up towards the buttocks, with knees kept close together to minimize drag. The feet rotate outward, with toes pointing out to prepare for the kick.

2. Propulsive: The legs kick outward and backward in a circular motion. The movement resembles a frog's kick. The soles of the feet press against the water to generate propulsion. The legs snap together quickly at the end of the kick, returning to the streamlined position. This snapping motion provides a burst of propulsion and minimizes drag.

Effective breaststroke relies on the precise timing and coordination of arm and leg movements. The typical sequence is:

1. Pull: The arms perform the outsweep and insweep.

2. Breathe: The head lifts for a quick breath during the insweep phase.

3. Kick: The legs execute the frog kick during or just after the arms begin the recovery phase.

4. Glide: A brief glide phase follows the kick, where the body remains streamlined.

Freestyle

The freestyle stroke technically means any stroke but butterfly, backstroke or breaststroke. Since the front crawl stroke is the fastest stroke, it is normally used as a synonym for freestyle. Freestyle is characterised by alternating arm movements and a flutter kick. The front crawl stroke can be divided into several distinct phases:

1. Entry: The swimmer extends one arm forward and enters the hand into the water, fingertips first.

2. Catch: Once the hand enters the water, the swimmer bends the elbow and initiates the catch phase. The hand and forearm act as a paddle, pushing water backward and downward to generate propulsion.

3. Pull: During this phase, the swimmer pulls the hand backward, past the hip, while maintaining a high elbow position.

4. Finish: As the arm reaches the end of the pull, the swimmer starts to rotate the body slightly, allowing the hand to exit the water near the thigh or hip.

5. Recovery: After the finish, the arm exits the water and starts to recover forward above the water's surface. The elbow bends, and the hand moves in a relaxed manner towards the entry position.

6. Entry and Extension: The recovering arm enters the water again, repeating the entry phase of the stroke. Simultaneously, the other arm, which has completed the pull phase, extends forward above the water, preparing for its entry.

Throughout the stroke, a flutter kick is used to provide additional propulsion and help maintain balance and body position. The legs execute a quick, continuous fluttering motion, with relaxed ankles and a small amplitude.

Individual Medley

The distance is divided into four equal parts that are swum in the sequence Butterfly, Backstroke, Breaststroke and Freestyle.

Drills

There are many drills for all strokes available. The creativity of the coaches is unlimited, and we can only cover some of them here. Please contact me if you would like your drill to be included in swiML. It is also unavoidable that the same drill is known under different names. When possible, synonyms for the drills are included in the list below. To better understand the strokes and the drills, I created a series of videos that you can watch on YouTube. Visit https://www.youtube.com/@christophbartneck, and look for the swiML playlist.

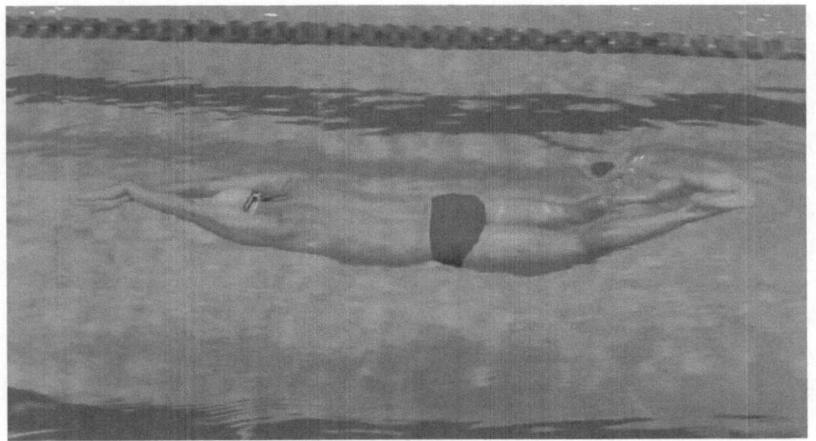

Figure 184: Demonstration videos are available on YouTube.

123

The 123 drill is often used for freestyle and backstroke. It is also known as the Double Tap drill. While one arm remains extended forward, the other arm starts the recovery phase. The swimmer is rotated so that the shoulder of the recovery phase arm is pointing upwards.

The swimmer moves the hand out of the water and taps on his/her shoulder before moving it back to where it exited the water. There, the swimmer taps the leg before moving the arm through the full recovery phase.

2Kick1Pull

This drill is normally used for breaststroke. The swimmer performs two kicks for every pull of the arms. The arms remain extended forward while the swimmer performs the additional kick.

2Pull1Kick

This drill is normally used for breaststroke. The swimmer performs two complete pulls with his/her arms for every kick. The legs remain extended backwards while the swimmer performs the additional pull.

3Kick1Pull

This drill is normally used for breaststroke. The swimmer performs three kicks for every pull of the arms. The arms remain extended forward while the swimmer performs the additional kicks.

3Pull1Kick

This drill is normally used for breaststroke. The swimmer performs three complete pulls with his/her arms for every kick. The legs remain extended backwards while the swimmer performs the additional pulls.

6KickDrill

This drill is suitable for freestyle and backstroke. After the initial glide, the swimmer transitions into a side-lying position. One arm remains extended in front, while the other is by the side, with the body rotated so that one shoulder points up and the other points down. In this position, the swimmer performs six kicks with the legs, maintaining a steady, rhythmic kick from the hips with minimal knee bend.

Following the six kicks, the swimmer rotates the body to the opposite side. The arm that was extended forward moves to the side, and the arm that was by the side extends forward. The swimmer then repeats the six kicks on the new side.

After completing the kicks on the second side, the swimmer rotates back to a prone initial position. The drill continues with the swimmer repeating the cycle of six kicks on each side.

This drill can be extended to 6,8,10, or 12 kick drills. This only changes the number of kicks between the arm strokes. This is used in swiML as 8KickDrill, 10KickDrill and 12KickDrill.

any

The swimmer can pick any of the drills.

bigDog

The Big Dog drill is used in freestyle. It is also known as the Long Dog drill. It is related to the Dog Paddle drill. In the recovery phase, the swimmer does not move his/her arm forward above the water, but under water.

dogPaddle

The best way to describe the Dog Paddle drill is to take inspiration from how dogs swim. The arms alternate in stroke. The swimmer extends his/her arms forward under water during the recovery phase. This is accompanied with a flutter kick.

fingerTrails

The Finger Trails drill is mainly used for the freestyle stroke. It is also known as the finger drag drill. The swimmer starts in the usual freestyle position. As the swimmer's hand exits the water at the end of the pull phase, instead of lifting the arm completely out of the water, he/she keeps the fingertips just above the surface.

The swimmer then lightly drags or "trails" their fingertips across the surface of the water as they move the arm forward in the recovery phase. This should be done smoothly and gently, ensuring that the fingertips maintain light contact with the water throughout the entire arm recovery.

fist

In the fist drill, the swimmer forms a fist instead of a flat extended hand while swimming.

other

By choosing the Other drill, the coach can specify unique drills by using the <instructionDescription> element.

scull

The Scull drill can be used swimming on the front or on the back. Let's start with the front scull. The swimmer has both arms extended to the front. The swimmer uses moves his/her hand and lower arm in a figure-eight "S" pattern. In the rear scull drill the swimmer is positioned on his/her back. The arms are positioned straight down the swimmer's sides and perform the sculling motion near their hips. The swimmer uses a gentle flutter kick to keep afloat and moving forward slowly.

singleArm

The Single Arm drill is suitable for butterfly, backstroke and freestyle. Instead of using both arms, the swimmer uses only one while the other remains extended towards the front or the back. The normal kick is used.

tarzan

The Tarzan drill is used in the freestyle stroke. The movements of the arms and legs are similar to normal freestyle. The swimmer keeps his/her head above the waterline. The swimmer holds his/her chin steady above the waterline without turning it left or right. The swimmer needs a strong flutter kick to maintain this position. The name for this drill is based on Johnny Weismuller's swimming style in the Tarzan movies (see Figure 128).

technic

This drill is suitable for all strokes. The swimmer focuses on the technic of the stroke. This normally means slowing down.

swiML Symbols

Figure 185 shows a quick reference list of symbols used in swiML.

Type	Symbol	Name
Structure	[	repetition
	|	continue
Length	100	length as a distance
	1 laps	length as laps
	1:00	length as time
Stroke	FL	butterfly stroke
	BK	backstroke stroke
	BR	breaststroke stroke
	FR	freestyle stroke
	IM	individual medley stroke
	IM Reverse	reverse IM stroke
	IM Overlap	IM overlap stroke
	IM Order	IM order stroke
	IM Reverse Order	reverse IM order stroke
	any	any stroke
	nr1	nr 1 stroke
	nr2	nr 2 stroke
	nr3	nr 3 stroke
	nr4	nr 4 stroke
	Not FL	not butterfly stroke
	Not BK	not backstroke stroke
	Not BR	not breaststroke stroke
	Not FR	not freestyle stroke
Drills	6KD	6 kick drill
	FT	finger trails
	123	123
	2K1P	2 kick one pull
	2P1K	2 pull one kick
Rest	☺	after stop rest
	←@_	since last rest
	@_	since start rest
	3 in 1 out	third in first out
Intensity	80%	percent effort
	♥80%	percent heart rate
	easy	effort zone
Extra	b	breath
	ꟿ	underwater

Figure 185: swiML symbols.

List of Figures

List of Listings

Bibliography

Rowland L Brooks, Cedric AB Smith, Arthur H Stone, and William T Tutte. The dissection of rectangles into squares. *Duke Mathematical Journal*, 7(1):312–340, 1940. DOI: 10.1215/S0012-7094-40-00718-9.

J. H. Conway. The weird and wonderful chemistry of audioactive decay. In Thomas M. Cover and B. Gopinath, editors, *Open Problems in Communication and Computation*, pages 173–188. Springer New York, New York, NY, 1987. ISBN 978-1-4612-4808-8. DOI: 10.1007/978-1-4612-4808-8_53.

Thomas Denes. *The Waterproof Swimmer: Swimming Workouts for Fitness Swimmers and Triathletes*. Ancient Mariner Aquatics, 2018. ISBN 9780965623018. URL https://search.worldcat.org/title/1231736915.

Clement Falbo. The golden ratio—a contrary viewpoint. *The College Mathematics Journal*, 36(2):123–134, 03 2005. DOI: 10.2307/30044835.

Erich Friedman. Packing unit squares in squares: A survey and new results. *The Electronic Journal of Combinatorics*, pages DS7–Aug, 2012. DOI: 10.37236/28.

Solomon W. Golomb. *Polyominoes*. Scribner, New York, 1965. ISBN 9780691024448. URL https://search.worldcat.org/title/982644.

Heinrich Heesch and Otto Kienzle. *Flächenschluss; System der Formen lückenlos aneinanderschliessender Flachteile*. Springer, Berlin„ 1963. ISBN 9783642948831. URL https://search.worldcat.org/title/1250086224.

Paul Kay and Willett Kempton. What is the Sapir-Whorf hypothesis? *American Anthropologist*, 86(1):65–79, 1984. DOI: 10.1525/aa.1984.86.1.02a00050.

Donald E Knuth. *The Art of Computer Programming: Fundamental Algorithms, Volume 1*. Addison-Wesley Professional, 1997. ISBN 9780201896831. URL https://search.worldcat.org/en/title/48246579.

Lubomir Markov. Heronian triangles whose areas are integer multiples of their perimeters. *Forum Geometriorum*, 7:129–135, 2007. URL https://forumgeom.fau.edu/FG2007volume7/FG200718.pdf.

George Orwell. *Nineteen Eighty-Four*. Secker and Warburg, 1949. ISBN 9780436350078. URL https://search.worldcat.org/title/1232 2711.

Henry J. Rogers. *The telegraph dictionary, and seamen's signal book, adapted to signals by flags or other semaphores; and arranged for secret correspondence, through Morse's electro-magnetic telegraph: for the use of commanders of vessels, merchants, &c.* F. Lucas, Jr Baltimore, Baltimore, 1845. URL https://www.worldcat.org/title/8925138.

Charles Safran. The Fibonacci Numbers. *Chance*, 5(1-2):43–46, 1992. DOI: 10.1080/09332480.1992.11882462.

M. Schofield and T. Griffith. *Plato: The Laws*. Cambridge Texts in the History of Political Thought. Cambridge University Press, 2016. ISBN 9780521859653. URL https://search.worldcat.org/en/title/9 23665304.

Laurence Sigler. *Fibonacci's Liber Abaci: A Translation into Modern English of Leonardo Pisano's Book of Calculation*. Springer Science & Business Media, 2002. ISBN 9780387954196. URL https://www.worldcat.o rg/title/48557588.

Lynne Truss. *Eats, Shoots & Leaves: The Zero Tolerance Approach to Punctuation*. Penguin, 2004. ISBN 9781861976123. URL https://se arch.worldcat.org/title/421779741.

J. Ziv and A. Lempel. Compression of individual sequences via variable-rate coding. *IEEE Transactions on Information Theory*, 24(5):530–536, 1978. DOI: 10.1109/TIT.1978.1055934.

Index